Extreme spiral grain with abrupt reversal of direction in *Hebe* sp.

John Maddern Harris

Spiral Grain and Wave Phenomena in Wood Formation

With 61 Figures

Springer-Verlag
Berlin Heidelberg New York
London Paris Tokyo

Dr. JOHN MADDERN HARRIS
12 Sumnervale Drive
Sumner
Christchurch 8
New Zealand

Series Editor:

T. E. TIMELL
State University of New York
College of Environmental
Science and Forestry
Syracuse, NY 13210, USA

Cover: Transverse section of *Pinus lambertiana* wood. Courtesy of Dr. Carl de Zeeuw, SUNY College of Environmental Science and Forestry, Syracuse, New York

ISBN 3-540-19382-0 Springer-Verlag Berlin Heidelberg New York
ISBN 0-387-19382-0 Springer-Verlag New York Berlin Heidelberg

Library of Congress Cataloging-in-Publication Data. Harris, John Maddern, 1921– . Spiral grain and wave phenomena in wood formation / John Maddern Harris. p. cm. – (Springer series in wood science) Includes index. 1. Spiral grain. 2. Wood – Anatomy. 3. Wood – Figure. I. Title. II. Series. QK647.H38 1988 674'.12–dc19 88-23477

Typesetting: K+V Fotosatz GmbH, Beerfelden; printing and binding: Brühlsche Universitätsdruckerei, Giessen
2131/3130-543210 – Printed on acid-free paper

Moreover, it is expedient that you know which are the veins and which the grain in timber, becaufe of the difficulty of working againft it: Thofe therefore are counted the veins which grow largeft, and are fofter, for the benefit of cleaving and hewing; *that* the grain, or pectines, which runs in waves, and makes the diverfe and beautiful chamfers which fome woods abound in to admiration. The Fir-tree, horizontally cut, has two circles of different fibres, which (when the timber comes to be cleft in the middle) feparate into four different waves, whence Pliny calls them *quadrifluvios*; and it is to be noted, that the nodous, and knotty part of thefe fort of trees, is that only which grows from the firft boughs to the fummit, or top, by Vitruvius termed the *Fufterna*, which both Baldus and Salmafius derive *à Fufte*. The other clean part, free of thefe boils, (being that which, when the fappy flab is cut away, is the beft) he calls *Sapinea*. Finally, the grain of Beech runs two contrary ways, and is therefore to be wrought accordingly; and indeed the grain of all timber ought well to be obferved; fince the more you work according to it, efpecially in cleaving, and the lefs you faw, the ftronger will be your work.

John Evelyn 1786 Silva: or a Discourse of Forest-trees and the Propagation of Timber in His Majesty's Dominions. Vol. 2, J. Todd, York.

Preface

It may seem that an entire book devoted to spiral grain is taking scientific specialisation to its extreme, but apart from its very considerable technical significance, spiral grain has also provided the inspiration for some of the most interesting and far-reaching recent research into plant anatomy, genetics, and physiology. As a wood scientist trying to assimilate the salient features of these complex disciplines, I am extremely grateful for the advice given so generously by many of the people whose names appear in the Acknowledgements. Nevertheless I accept full responsibility for any errors or serious omissions.

I have tried in each chapter to include sufficient background information on the relevant specialist disciplines to enable the reader to understand the significance and direction of current research without recourse to other texts. In doing so, I am aware that many topics have had to be abbreviated to the point where they may seem inadequate to the expert. Elsewhere, as in the discussion of growth substances for example, I have been deliberately selective in order to guide the reader along what appear to be the main routes through a particular topic. Future research may well show that some of these routes are dead ends, and that what appear to be side tracks today provide the way to true enlightenment. But these are the risks normal to scientific judgment.

As in all branches of science, older observations can seem almost comical in the light of current knowledge, yet many of these early concepts may hold a grain of truth that can be used to start anew. For this reason I have reviewed all those papers available to me that were written since the turn of the century, and have been selective only with respect to works dated before 1900. The conclusions listed at the end of each chapter provide a more personal view of those results that seem most significant, and of future directions that seem most promising.

October 1988

John Maddern Harris
Christchurch, New Zealand

Acknowledgements

I am indebted to many of my former colleagues at the New Zealand Forest Research Institute in Rotorua for their encouragement and advice, particularly the Director, Dr. J.A. Kininmonth, I.A. Andrew, R.D. Burdon, M.J. Collins, D.J. Cown, Mrs J. Griffith, D.A. Rook, D.R. Smith, G.B. Walford, and D. Whitehead. Messrs. D. Stonehouse and G.D. Young provided the information and communication essential for me to write in absentia, and Mrs. L. Ernst kindly typed the manuscript.

It is also a pleasure to acknowledge the assistance of Prof. W.L. Jones, Dr. B.G. Butterfield, Dr. J.F.C. Walker, and Dr. A.G.D. Whyte of the University of Canterbury, Mr. R.N. Patel of Botany Division, DSIR Lincoln, Dr. J.A. Romberger, and Dr. Jane Hill of the Plant Physiology Unit, Beltsville, Maryland, USA, and Prof. T. Sachs of the Hebrew University of Jerusalem in Israel.

The outstanding contributions to the topic of spiral grain by scientists at the universities of Wroclaw, Silesia, and Warsaw in Poland are widely acknowledged, and I am grateful to all those who sent me copies of their papers. In particular, it was a special pleasure to me that I had the opportunity to discuss this important body of work, in person and by correspondence, with Dr. Beata Zagórska-Marek of Wroclaw University.

October 1988

John Maddern Harris
Christchurch, New Zealand

Contents

Chapter 1

Defining Spiral Grain

1.1 Patterns of Spirality

Spiral grain: a fundamentally simple concept in which wood grain – as seen on the outside of a log, for example – runs around it in a spiral rather than parallel with the axis. Even the most casual observer will have seen indications of spirality in the cracks running around old dead trees, or around well-weathered transmission poles, fence posts, and the like. On the face of it a minor curiosity and hardly the subject for extensive study, one might think. Yet so many technical problems arise in using roundwood, sawn timber, and veneers made from trees containing spiral grain that some wood scientists have taken the view that this is the most serious single defect in various plantation-grown softwoods (Rault and Marsh 1952, Banks 1953, 1969, de Villiers 1974).

The technical disadvantages arising from using spirally grained timber are certainly major reasons why so much interest has been expressed on the subject, but numerous workers have also seen spirality as being just one manifestation of the many ways in which wood grain may deviate from strict polarity. They take the view that other (often highly prized) features of wood, such as interlocked grain and wavy grain, arise in the same way anatomically, and must therefore be regarded as part of the same phenomenon.

In this broader sense the family of grain deviations, of which simple spirality is just one member, can either render sawn timber almost worthless, or enhance its value enormously. As a result, foresters, wood anatomists, geneticists, and physiologists have all been induced to look for the causes of these grain patterns and, as often happens in studies of apparent "aberrations", their work has greatly advanced our appreciation of the mechanisms that control "normal" growth and development.

These issues will be taken up in later chapters, but the basic question still remains as to how spiral grain may be defined? What is the justification for including such phenomena as interlocked grain and wavy grain in this general category? Indeed, we must also ask, the others have (Priestley 1945, Vité 1967), whether all instances of even apparently simple spirals in stems of different tree species are the same or whether the implied unification is only semantic.

In this chapter these questions will be addressed not to arrive at a definitive answer, but only to try to discover which of the many curious, ornamental, or bizarre grain patterns encountered in wood may be able to contribute to our understanding of spirality, and which of them may be deemed, on present evidence, to be unrelated.

In due course it will become apparent that a simple spiral, i.e. one that is completely consistent in its direction and degree throughout a stem, is seldom if ever encountered in nature. Where such simple spirals have been described they are usually found to refer strictly to the external appearance of a short stem section such as a log. Nevertheless it will be convenient to use the simple spiral as a starting point for more detailed considerations.

1.1.1 The Simple Spiral

A spiral of given diameter can be defined in terms of its direction of rotation and the angle it makes with the axis of the cylinder which it encompasses. Both of these features have been described in various ways in the past but not without a certain degree of confusion.

1.1.1.1 Direction of the Spiral

It is current usage to define the direction of spirality as being left-hand (LH) spiral grain when the deviation is to the left of the upper extremity of the longitudinal axis of a tree as viewed by an observer on the ground, and right-hand (RH) spiral grain when the deviation is to the right. In the same sense left grain angle and right grain angle are expressions used to describe the degree of deviation. Some writers have also applied the letters S and Z to grain direction, especially for indicating cellular orientation in anatomical studies, the direction of the oblique stroke in either letter being used to indicate left and right inclination respectively.

Numerous other expressions have been used to describe spiral direction, reflecting no doubt the notorious difficulty of so doing without waving of arms and twisting of wrists. In the German literature the terms "sonnig" and "widersonnig" are frequently encountered. Their meaning is "with the sun" and "against the sun", and hence (in the Northern Hemisphere!) they relate to LH and RH spiral grain respectively. The English terms clockwise and counterclockwise have been similarly applied from the viewpoint of an observer looking down the stem from above, and Misra (1939) pointed out that this probably derives from the system used to describe the cutting of screw threads.

In his delightful dissertation on traditional and literary observations of spiral grain, Meyer (1949) traced back the concept of left-handedness contained in the expressions sinistrorse, sinistral and sennotes, or right-handedness as dextrorse, dextral, and dextres, from their earliest beginnings. Starting from the ancient Greek (originating from Sanskrit) through Latin and later through Italian, French, and English, he gave numerous examples from classical and more recent literature.

Unfortunately, despite the extensive historical background for the concept, misunderstandings have arisen in the past, and it is necessary to be aware of these when comparing the results given by different authors. One obvious danger arises in using a concept such as clockwise and counterclockwise if the observer looks at the rotation from the bottom to the top from a viewpoint below the base or above the apex, when the directional sense becomes reversed, as it was by Jacobs (1935).

The extensive data of Braun (1854) which were used by the authors or compilers of most of the standard 19th century textbooks of botany and forestry (Champion 1924) also indicated spiral directions which were the reverse of those in current usage. This arose because the German word "Drehwuchs" meaning "turning growth" led to the concept that what we now call LH spiral grain was "turning" to the right as it ascended the stem. Thus "Rechtsdrehung" (right-turning) in these early accounts refers to a LH spiral as defined above, and more recent writers (e.g., Löhr 1966) have persisted in referring to it as an alternative which is more appropriate "in the botanical sense".

From the beginning observers noticed that some species such as *Betula pendula* are almost invariably straight grained, whereas in others LH spiral grain (e.g., *Malus domestica*) or RH spiral grain (e.g., *Pyrus communis*) is strongly dominant (Burger 1946).

1.1.1.2 Degree of Spirality

It is customary to measure grain angle as a deviation from the longitudinal axis, and the methods used for measurement will be described in Chapter 2. However, in a few instances "angle of ascent" (German "Steigungswinkel") has also been used, in which case the values given are the complement of the grain angle whether to the left or the right.

Some species have become notorious for producing extreme spiral grain. *Pinus roxburghii,* the chir pine of India, has attracted much attention on this account both in its native habitat (Canning 1915, Champion 1924, 1927a, 1930, 1931, 1945, Kadambi and Dabral 1955) and wherever it has been grown as an exotic (Banks 1969, Nikles and Smith 1974). Champion (1924) provided photographs showing how extreme LH spiral grain may approach the horizontal in the basal region of indigenous chir pine stems. Smythies (1915) gave a colourful description: "In such trees the twist is carried into the branches, and they frequently present the most bizarre and fantastic contortions, both the main stem and the branches forming amazing knots. One might imagine the trees were writhing in perpetual and awful agony."

In his original observations Braun (1854) described some extreme examples of spiral grain in hardwood species including angles up to 45° in *Punica granatum* and angles nearly as great in *Sorbus aucuparia, Syringa vulgaris,* and *Aesculus hippocastanum.* Hartig (1895) recorded one log of *Pinus sylvestris* in which the spiral angle approached 90°, and Jones (1963) illustrated a similar example in a section of *Pseudotsuga menziesii* less than 100 mm diameter. Figure 1.1 shows extreme spiral grain, approching 90° in places, on the outside of a fallen branch of *Pinus sylvestris.*

1.1.1.3 Defect or Normal Growth?

Extreme spirality, approaching the condition of horizontal grain, is not only almost useless for anything but reconstituted wood products, but also poses many

Fig. 1.1. Severe spiral grain on the outside of a fallen branch of *Pinus sylvestris.* (Photo T. E. Timell)

questions in wood anatomy and tree physiology (e.g., how water is conducted along the stem). These issues will be discussed in later chapters, but in many respects such extreme examples have the appearance of a growth pattern that has gone out of control, and may be regarded as deviations from normal growth and development. As in most biological phenomena it is nevertheless often difficult to draw the line between what is normal and what is defect.

Exceptions are frequently encountered to all apparent "rules" of spirality, and the literature as a whole makes confusing reading with its mass of apparently contradictory statements. No sooner is the spiral pattern within a species firmly established, such as the reliably straight grain of *Betula pendula* (grain angle seldom more than 3°), or the inevitability of RH spiral grain where it occurs in *Aesculus hippocastanum,* than exceptions are reported (e.g., Löhr 1966).

This is one reason why many writers have sought safety in numbers and have reported observations of hundreds, sometimes thousands, of trees. Table 1.1 provides an historical sequence of published results based on large numbers of trees. Some of these are naive observations of amateur naturalists, intrigued by the varied patterns they see revealed on old dead tree trunks after fire or other natural disasters, and some are the painstaking efforts of scientists who have accepted the need for a sound numerical basis for genetic and other studies.

In their entirety they clearly indicate that some degree of spiral grain is by no means an abnormal feature of wood formation in trees. For wood products used by man spiral grain may well be a "defect", but if it were a defect to tree development and survival it is hard to believe that the processes of natural selection would not have eliminated it by now, or at least reduced it to an occasional feature en-

Table 1.1. A historical review of numerically major studies of spiral grain

Author(s)	Date	Species	Numbers examined	Comments
Braun	1854	167 spp.	? 1500	111 spp. showed spiral grain
McCarthy and Hoyle	1918	*Picea rubens*	2372	48% had spiral >1 in 20
Champion	1924	*Pinus roxburghii*	4200	Seedlings with LH spiral
			1400	4-year-old trees
			700	30-year-old trees
			? 4000	Older trees in forest stands
Cahn	1931	*Thuja plicata*	312	More than half RH spiral
Jacot	1931	*Thuja orientalis*	438	More than half LH spiral
		Juniperus chinensis	38	More than half LH spiral
Koehler	1931	*Abies lasiocarpa*	396	85% RH; 14% LH; 1% straight
		Pseudotsuga menziesii	463	20% LH; 2% RH; 78% straight
Went-worth	1931	*Pinus* spp. (high alt.)	400	96% RH; 4% LH
Herrick	1932	*Pinus palustris*	1527	53% RH; 24% LH; 23% straight
Knorr	1932	*Pinus banksiana*	486	Of spiralled trees 56% RH; 44% LH
		Mixed: (California)	822	13% RH; 14% LH; 73% straight
Burger	1941	*Picea abies, Abies alba*	3300	LH in youth; tending RH in age
	1946	*Pyrus communis*	3843	Increasingly RH with age
		Malus domestica	3446	Increasingly LH with age
	1947	*Quercus robur*	2195	Predominantly RH with age
		Quercus petraea	2225	Predominantly RH with age
	1950a	*Fagus sylvatica*	1808	Mainly straight: RH more frequent
	1950b	*Larix decidua*	1774	LH in youth; RH with age
		Picea abies (high alt.)	1117	Similar to med. alt. (Burger 1941)
Rault and Marsh	1952	*Pinus roxburghii*	300 plots	Genetics, age, growth rate implicated

Table 1.1 (continued)

Author(s)	Date	Species	Numbers examined	Comments
Rao	1954	*Cedrela toona*	484	15% RH; 15% LH; 70% straight
Northcott	1957	12 spp. old trees 17 spp. seedlings	594 709	99% showed some spiral 81% showed some spiral
Ohkura	1958	30 Japanese conifer spp. 34 O'seas conifer spp. 67 Hardwood spp.	150 65 390	Comparison N and S Hemispheres
Kano et al.	1964	*Larix kaempferi*	200	Predominantly LH in youth
Gerischer and Kromhout	1964	*Pinus patula*	2400	Selection for breeding
Lowery	1966	*Pseudotsuga menziesii*	775	45% LH; 48% RH; 7% straight
		Pinus contorta	113	30% LH; 69% RH; 1% straight
		Larix occidentalis	97	77% LH; 19% RH; 4% straight
		Picea engelmannii	104	59% LH; 37% RH; 5% straight
Löhr	1966	*Aesculus hippocastanum*	388	Straight 69%; RH 31%
Paterson	1967	*Cupressus* spp. *Pinus patula* *Pinus radiata*	231 235 157	Assessment for breeding
Krempl	1970	*Picea abies*	120	Mainly LH turning RH in age
Mikami and Nagasaka	1975	*Larix kaempferi*	358	Predominantly LH in youth
Kostov and Todorov	1977	*Quercus petraea* *Quercus conferta*	287	25% logs downgraded through spiral

countered in isolated trees not subject to the competition of better-regulated neighbours. In practice the opposite is true. Spiral grain has been detected in fossil trees of great antiquity and has thus been a persistent feature for many millions of years (Champion 1924). Moreover, it is not uncommon to find trees that have developed the most extreme degree of spirality, in *Pinus roxburghii* for example, co-existing without apparent difficulty alongside straight-grained trees in dense stands of high forest. If extreme spiral grain presents conceptual difficulties to the scientific mind, the trees themselves clearly have little difficulty in adapting to it. Indeed, its frequency and strong development in some species in particular areas have raised the question of possible competitive advantages (Sect. 4.5).

On the other hand, there are circumstances which suggest that spirality sometimes reflects a breakdown in the strict control of polarity within stems. For example, in the development of localised or sometimes quite extensive oblique grain patterns after injury, or in the frequently observed increase of spiral angle during tree senescence when cessation of active crown growth may bring about many metabolic changes. Braun (1854) acknowledged the presence of spiral grain in certain genetically abnormal individuals which he termed "Monstrosität", yet he set these aside as being different from the general run of spiral grain in trees. In common with many later writers, notably Mayer-Wegelin (1956) and Northcott (1957), Braun came to the conclusion that spiral grain is not a sickness but a normal attribute for many trees.

For the time being we need only note that spiral grain may not always arise in the same way, nor have the same consequences for tree development and survival. Our present purpose is only to decide which of these phenomena are "related" insofar as a clearer understanding of one may help to explain others.

1.1.2 Change in Spiral Angle and Direction

As explained above, many writers have simply recorded the external appearance of standing trees. Some have observed that trees of small diameter tend to have a different spiral direction from those of larger diameter, and some have related this to tree age. However, not all have appreciated that both spiral angle and direction may change markedly throughout the life of a tree, though this fact was clearly recorded by the earliest writers (Braun 1854, Hartig 1895).

Changes in grain angle were also obvious enough to the craftsmen who worked in wood, though these were not usually the people who wrote technical papers, demonstrating a regrettable lack of communication between the two groups which unfortunately is not unknown even today. In Europe, for example, craftsmen such as shinglemakers and coopers who split rather than sawed their basic work pieces from the log, knew full well that logs which were straight grained or had only a small LH spiral would split better than those that had RH spiral (Burger 1941). This is because the species they were using commonly developed LH spirality in youth, which in some trees developed into extreme LH spirality or reversed to RH spirality later in life. Logs with moderate LH spirality could be relied upon to produce the more consistent grain angle required for successful splitting.

It is difficult to generalise about stem patterns of spiral grain, but Noskowiak (1959, 1963) made the point that one pattern occurs very frequently in many softwoods and that, although variability is even greater in hardwoods, there is a tendency for the pattern within them to be the reverse of that in softwoods.

1.1.2.1 Stem Pattern in Softwoods

In simplest terms the wood immediately adjacent to the pith in softwoods is straight grained, but LH spiral develops very soon and continues to increase during subsequent growth, often reaching a maximum value within the first ten annual growth layers. Thereafter the grain angle decreases towards zero, and later in the life of the tree it may change direction to RH with grain angles in this direction increasing slowly for as long as stem growth continues.

The initiation of spirality in primary xylem and first-formed secondary xylem has been frequently implied by the use of phrases such as "high initial spirality" and so on. It may well arise in some cases from direct association with the phyllotaxic spiral (Northcott 1957), or by twisting growth in the leading shoot. In fact Jablanczy (1969) described nearly 45° of twisting in the hypocotyl of a seedling of *Picea abies,* and it is certainly difficult to conceive of spirality starting earlier that that.

However, close examination of many species in which a considerable degree of spirality may be developed by the end of the first year's growth has shown that wood immediately adjacent to the pith is straight grained (Ohkura 1958, Brazier 1965, Harris 1965, Krempl 1970, Mikami 1973, Ozawa 1973, Kawaguchi et al. 1979). Even seedlings of *Pinus roxburghii* destined to produce severe spiral grain in later growth were reported by Champion (1924) to be straight grained at the end of the first year's growth. "High initial spirality" often proves on closer examination to refer to grain angles at the second or third annual growth layer from the pith (e.g., Kromhout and Toon 1977). This reflects the difficulty of dissecting out and measuring grain angles in small growth layers with a high degree of curvature. In addition, the presence of needle traces (xylary connections to the stem leaves) causes stem tracheids close to the pith to follow a more or less sinuous path between the phyllotaxic spirals, which complicates any measurement of grain angle even when the overall direction does not deviate consistently from that of the stem axis.

Once spiral grain has been initiated, the time taken to reach maximum LH spirality is extremely variable both within and between species, as is the time taken to cross over to RH spirality. Examples from three species which have been intensively investigated [*Pseudotsuga menziesii, Larix kaempferi* (syn. *L. leptolepis*), and *Pinus radiata*] will serve to illustrate these features.

Northcott (1957) described *Pseudotsuga menziesii* from ten different localities in which the average time taken to reach maximum LH spirality varied from 20 to more than 200 years, and the average time taken to change over to RH spirality varied from 100 to 300 years. Of the trees examined, 64% followed the typical pattern of LH spiral in youth changing to RH spiral in age, 19% had continuously increasing LH spirality or failed to reduce the maximum value significantly, and 17% "exhibited most of the other possible combinations".

In *Larix kaempferi* maximum LH spirality occurs between the second and ninth annual growth layers from the pith more than 80% of the time (Ozawa 1972). The change from LH to RH spirality is found in 53% of the population and occurs between the fifth and fortieth annual growth layers. Forty-one percent of the trees retain LH spirality up to 40 years and 6% are more or less straight grained (Nakagawa 1972, Mikami 1973).

In *Pinus radiata* the maximum value is often reached by the outside of the second or third annual growth layer, although values may increase for 12 years or more (Chattaway 1959). The cross-over from LH to RH spirality may happen as early as eight growth layers from the pith but some trees may retain LH spirality at least to the age of 50 years, the age at the time of clearfelling usually being considerably less than that for *P. radiata.*

The consistency of patterns of spirality developed at different heights within individual stems will be dealt with in Chapter 2 because it obviously has a bearing on the reliability with which patterns measured at one stem cross-section can be applied to the stem as a whole. At this point it need only be said that whereas grain patterns may prove to be sufficiently consistent to permit extrapolation in some species, especially amongst plantation-grown softwoods, it is a feature which should never be taken for granted.

An obvious exception is to be found amongst the tropical pines. Although research into these species is relatively recent, there is already considerable evidence that spiral grain can vary in a highly unpredictable way in *Pinus caribaea* (Wiggins 1979, Whyte et al. 1980, Cown et al. 1983), *P. kesiya* (Burley and Andrew 1970), *P. oocarpa* (Houkal 1982), and *P. merkusii* (Andrew and Burley 1973). These conclusions would be less significant if the grain angles encountered were small, but unfortunately this is not the case. Results presented by Cown et al. (1983) are fairly typical: (1) 79% of all angles measured were LH spirals; (2) 28% of all angles exceeded 5° and 3% exceeded 10°; (3) no consistent radial trend was apparent at any height in the stem; (4) development of spirality differed markedly at different heights within individual stems. Unfortunately many assessments of spiral grain have been based on a single sample taken from each tree at breast height (b.h. = 1.4 m). Results have even been used to exclude otherwise superior trees from breeding programmes (e.g., Smith 1973, de Villiers 1974).

Other exceptions to the common stem pattern for softwoods are those species in which the pattern is reversed, i.e., in which the initial spiral is RH and reversal to LH may occur in older stems. Ohkura (1958) pointed out that these are by no means uncommon amongst Japanese species including the *Pinus* species *P. koraiensis, P. parviflora,* and *P. pumila.* Other Japanese species with initial RH spiral are *Podocarpus macrophylla, Cryptomeria japonica, Thuja standishii,* and *Thujopsis dolabrata.* Ohkura's results from countries outside Japan led him to suggest that softwoods with initial RH spirality are more frequent in the Southern Hemisphere than in the Northern. This observation was supported by Smith and Smart (1972) who reported that RH spiral is found in most indigenous Australian softwoods. However, neither implies any biological significance in this geographical bias. (To answer a question often asked by scientists from the Northern Hemisphere, it may be appropriate at this point to assure them that northern trees do *not* reverse their initial spiral grain direction when grown below the equator!)

Table 1.2 summarises some of the more significant results (based on not less than three samples) for stem patterns of spiral grain in softwoods. Where many reports deal with a single species some selection has been made to allocate priority, or to extend geographical coverage and so on. Those species in which more than one reversal of grain direction have been recorded are marked Int., denoting interlocked grain as described in Section 1.2.

Table 1.2. Predominant spiral patterns in softwoods

Species	Reference	Pattern[a]	Remarks
Abies alba	Burger 1941	L – R	
Abies firma	Ohkura 1958	L – R	
Abies homolepis	Ohkura 1958	L – R	Int.
Abies mariesii	Ohkura 1958	L – R	
Abies pectinata	Braun 1854	L – R	
Abies sachalinensis	Ohkura 1958	L – R	Int
Abies veitchii	Ohkura 1958	L – R	
Agathis palmerstonii	Balodis 1971	R – L	
Araucaria cunninghamii	Balodis 1971	R – L	
Cephalotaxus drupacea	Ohkura 1958	L – R	
Chamaecyparis obtusa	Ohkura 1958	O	Variable
Chamaecyparis pisifera	Ohkura 1958	O	Variable
Cryptomeria japonica	Ohkura 1958	R	
Ginkgo biloba	Ohkura 1958	L	Int.
Juniperus communis	Braun 1854	R	– L in age?
Juniperus rigida	Ohkura 1958	L	
Juniperus virginiana	Braun 1854	R	
Larix decidua	Burger 1950b	L – R	
Larix kaempferi	Ohkura 1958	L – R	
	Nakagawa 1972	L – R	
	Ozawa 1973	L – R	
	Kawaguchi et al. 1979	L – R	
Larix occidentalis	Lowery 1967a, b	L – R	
Picea abies	Braun 1854	L – R	
	Burger 1941	L – R	
	Krempl 1970	L – R	
Picea glehnii	Ohkura 1958	L – R	
Picea hondoensis	Ohkura 1958	L – R	
Picea jezoensis	Ohkura 1958	L – R	
Pinus austriaca	Braun 1854	L – R	
Pinus canariensis	Banks 1969	L – O	Young trees
Pinus caribaea	Whyte et al. 1980	L	Very irregular
	Cown et al. 1983	L	Very irregular
Pinus contorta	Northcott 1957	L – R	
Pinus densiflora	Ohkura 1958	L – R	
Pinus elliottii	Banks 1969	WL	
	Balodis 1971	L – R	
Pinus koraiensis	Ohkura 1958	R – L	
Pinus merkusii	Andrew and Burley 1973	L	Very irregular
Pinus oocarpa	Houkal 1982	L – R	Very irregular
Pinus parviflora	Ohkura 1958	R – L	
Pinus patula	Banks 1969	L – O	Young trees
Pinus pinaster	Nicholls 1967b	L – O	
Pinus ponderosa	Paul 1956	L – O or WR	
Pinus pseudostrobus	Nicholls 1967b	L – O	
Pinus pumila	Ohkura 1958	R – L	
Pinus radiata	Jacobs 1935	L – R	
	Chattaway 1959	L – R	
	Harris 1965	L – R	Int.
	Fielding 1967	L – R	
Pinus roxburghii	Champion 1924	L – R	Extreme angles
	Rault and Marsh 1952	L increasing	
	Banks 1969	L increasing	
	Nikles and Smith 1974	L increasing	

Table 1.2 (continued)

Species	Reference	Pattern[a]	Remarks
Pinus sylvestris	Braun 1854	L – R	
	Hartig 1895	L – R	
Pinus taeda	Banks 1969	L – O	Young trees
	Balodis 1971	L – R	
Pinus thunbergii	Ohkura 1958	L – R	
Podocarpus macrophylla	Ohkura 1958	R	
Podocarpus nagi	Ohkura 1958	L – R	
Pseudotsuga menziesii	Northcott 1957	L – R	Int.
	Elliott 1958	L – R	
Taxus cuspidata	Ohkura 1958	L	
Thuja standishii	Ohkura 1958	R – L	
Thujopsis dolobrata	Ohkura 1958	R – L	
Torreya nucifera	Ohkura 1958	L – R	
Tsuga diversifolia	Ohkura 1958	L – R	
Tsuga heterophylla	Kennedy and Elliott 1957	L – R	
	Elliott 1958	L – R	
Tsuga sieboldii	Ohkura 1958	L – R	Int.

[a] O = straight grained or less than 2°; L = LH spiral; R = RH spiral; W = weak; Int. = interlocked, more than one reversal of direction recorded. Note: Initial spiral direction refers to that *developed* first, and not at the pith.

1.1.2.2 Stem Pattern in Hardwoods

Grain patterns in hardwoods are both less predictable and more complex than those in softwoods. In particular the intricate patterns of interlocked and wavy grain dealt with in Sections 1.2 and 1.3 are much more frequently encountered in hardwoods. In Noskowiak's (1959) self-proclaimed "cautious statement" that the stem pattern in hardwoods tended to be the reverse of that in softwoods, he admitted that it was based on very few studies.

By far the greatest additional contribution has come from one source – Ohkura (1958) described patterns of split log cross-sections for more than 140 species of hardwoods. These and other results are summarised in Table 1.3. In this table only information based on more than two samples per species has been included, and species are listed as straight grained in which the majority of samples have grain angles less than 2°. It can be seen that change of spiral grain direction (i.e., excluding interlocked grain) is encountered less frequently in hardwoods than in softwoods, and that it is hard to generalise about spiral grain in hardwoods at all.

However, Table 1.3 is not representative of all hardwoods because not all have been studied equally. Many commercial hardwoods have the reputation of being straight grained, and this is probably one reason why spirality has not been investigated as extensively as in softwoods. Studies that have been done using large numbers of stems often suggest that spiral grain has more of the aspect of an "aberration" in hardwoods, as compared with softwoods in which it often appears as the normal, and indeed inevitable, developmental pattern. Thus Burger's

Table 1.3. Spiral and interlocked grain in hardwoods

Species	Reference	Pattern[a]	Remarks
Acanthopanax spinosus	Ohkura 1958	O	
Acer carpinifolium	Ohkura 1958	L or R	
Acer pictum	Ohkura 1958	O	
Acer rufinerva	Ohkura 1958	O	
Acuba japonica	Ohkura 1958	O	
Aesculus hippocastanum	Braun 1854	R	
Aesculus turbinata	Ohkura 1958	O – R	
Afrormosia elata	Scheiber 1973	Int.	Regular
Alnus firma	Ohkura 1958	R	
Alnus glutinosa	Braun 1854	O	
Alnus rubra	Kennedy and Elliott 1957	R – L	
Alnus tinctoria	Ohkura 1958	O or WR	
Angophora spp.	Bootle 1983	Int.	
Anisoptera cochinchinensis	Scheiber 1973	Int.	
Anogeissus acuminata	Limaye 1954	Int.	
Antiaria spp.	Scheiber 1973	Int.	Regular
Apuleia praecox	Bootle 1983	Int.	
Aucoumea klaineana	Scheiber 1973	Int.	
Backhousia hughesii	Bootle 1983	Int.	
Betula ermani	Ohkura 1958	R	
Betula pendula	Braun 1854	O	
Calophyllum spp.	Bootle 1983	Int.	
Canarium schweinfurthii	Scheiber 1973	Int.	
Castanea sativa	Ohkura 1958	O	
Cedrela toona	Limaye 1954	Int.	
Ceiba pentandra	Scheiber 1973	Int.	
Cercidiphyllum japonicum	Ohkura 1958	R	
Cercidiphyllum magnificum	Ohkura 1958	R	
Chlorophora excelsa	Bootle 1983	Int.	
Chukrasia spp.	Scheiber 1973	Int.	Regular
Clethra barvinervis	Ohkura 1958	R	
Cornus controversa	Ohkura 1958	O – R	
Corylus heterophylla	Ohkura 1958	L	
Cratoxylon arborescens	Bootle 1983	Int.	
Dacryodes buettneri	Scheiber 1973	Int.	
Dacryodes pubescens	Scheiber 1973	Int.	
Dalbergia latifolia	Scheiber 1973	Int.	Regular
Daniella klainei	Détienne 1979	Int.	Annual
Daniella sayauxii	Détienne 1979	Int.	Annual
Daniella thurifera	Détienne 1979	Int.	Annual
Diospyros kaki	Ohkura 1958	O	
Diospyros lotus	Ohkura 1958	O – R	
Distemonanthus benthamianus	Scheiber 1973	Int.	
Dracontomelum dao	Scheiber 1973	Int.	
Entandrophragma angolense	Scheiber 1973	Int.	
Entandrophragma candollei	Scheiber 1973	Int.	Regular
Entandrophragma cylindricum	Scheiber 1973	Int.	Regular
Entandrophragma utile	Scheiber 1973	Int.	
Erythrophleum spp.	Scheiber 1973	Int.	Regular
Eucalyptus spp.	Bootle 1983	Int.	
(34 spp. listed as frequently interlocked, and others occasionally)			
Euonymus macropterus	Ohkura 1958	L or R	
Euonymus sieboldianus	Ohkura 1958	L or R	

Table 1.3 (continued)

Species	Reference	Pattern[a]	Remarks
Euonymus striatus	Ohkura 1958	L or R	
Fagus crenata	Ohkura 1958	O	
Fagus sylvatica	Burger 1941	O – WR or L	
Flindersia australis	Bootle 1983	Int.	
Flindersia brayleyana	Bootle 1983	Int.	
Fraxinus sieboldiana	Ohkura 1958	O	
Gmelina spp.	Bootle 1983	Int.	
Gleditsia japonica	Ohkura 1958	O – WR	
Gossweilerodendron balsamiferum	Scheiber 1973	Int.	
Guarea spp.	Scheiber 1973	Int.	
Guibourtia spp.	Scheiber 1973	Int.	
Guilandina echinata	Scheiber 1973	Int.	
Ilex macropoda	Ohkura 1958	R	
Ilex pedunculosa	Ohkura 1958	R	
Ilex rotunda	Ohkura 1958	R	
Intsia bijuga	Bootle 1983	Int.	
Juglans regia	Braun 1854	O	
Juglans sieboldiana	Ohkura 1958	O	
Kalopanax pictum	Ohkura 1958	O	
Khaya spp.	Scheiber 1973	Int.	
Koompassia spp.	Bootle 1983	Int.	
Ligustrum obtusifolium	Ohkura 1958	L	
Liquidambar formosana	Harzmann 1965	Int.	
Liquidambar styraciflua	Webb 1967	Int.	
Lindera obtusifolia	Ohkura 1958	L or R	
Lovoa spp.	Scheiber 1973	Int.	
Lyomia ovalifolia	Ohkura 1958	O – R	
Magnolia obus	Ohkura 1958	L	
Magnolia obovata	Ohkura 1958	L	
Malus domestica	Burger 1946	R – L	
Malus toringo	Ohkura 1958	L – R	
Mesua ferrea	Kadambi 1951	Int.	
Michaelia excelsa	Limaye 1954	Int.	
Micromeles alnifolia	Ohkura 1958	L	
Milletia laurentii	Scheiber 1973	Int.	Regular
Nesogordonia papaverifera	Scheiber 1973	Int.	Regular
Nyssa sylvatica	Krawczyszyn and Romberger 1980	Int.	
Octomeles sumatrana	Bootle 1983	Int.	
Oxystigma oxyphyllum	Scheiber 1973	Int.	
Palaquium ellipticum	Kadambi 1951	Int.	
Palaquium spp.	Bootle 1983	Int.	
Parabezoin praecox	Ohkura 1958	O – R	
Parashorea spp.	Bootle 1983	Int.	
Pittosporum tobira	Ohkura 1958	O	
Platanus acerifolia	Krawczyszyn 1971	Int.	
Platanus occidentalis	Krawczyszyn 1971	Int.	
Poeciloneuron indicum	Kadambi 1951	Int.	
Populus trichocarpa	Northcott 1957	WR – R	
Populus sieboldiana	Ohkura 1958	R	
Pourthiaea villosa	Ohkura 1958	L – R	
Prunus grayana	Ohkura 1958	O – WR	

Table 1.3 (continued)

Species	Reference	Pattern[a]	Remarks
Prunus subhirtella	Ohkura 1958	O – WR	
Pterocarpus macrocarpus	Bootle 1983	Int.	
Pterocarpus pedatus	Scheiber 1973	Int.	Regular
Pterocarpus soyauxii	Scheiber 1973	Int.	Regular
Pterocarpus corymbosa	Ohkura 1958	O – R	
Pyrus communis	Burger 1946	WR – R	
Quercus acutissima	Ohkura 1958	O – WR	
Quercus crispula	Ohkura 1958	O – WR	
Quercus mongolica	Ohkura 1958	O – WR	
Quercus robur	Braun 1854	R	
Quercus serrata	Ohkura 1958	O – R	
Quercus sessiliflora	Burger 1947	O – R	
Robinia pseudoacacia	Ohkura 1958	O – R	
Rhododendron degronianum	Ohkura 1958	O – R	
Rhododendron japonicum	Ohkura 1958	O	
Salix bakko	Ohkura 1958	O or WR	
Salix integra	Ohkura 1958	O or WR	
Sambucus nigra	Braun 1854	L or R	
Sambucus sieboldina	Ohkura 1958	WL or WR	
Shorea albida	Bootle 1983	Int.	
Shorea robusta	Limaye 1954	Int.	
Sloanea australis	Bootle 1983	Int.	
Styrax japonicum	Ohkura 1958	L or WR	
Syncarpia glomulifera	Bootle 1983	Int.	
Syncarpia hillii	Bootle 1983	Int.	
Syzygium spp.	Bootle 1983	Int.	
Tariettia spp.	Scheiber 1973	Int.	
Tiephemella africana	Scheiber 1973	Int.	
Tiephemella heckellii	Scheiber 1973	Int.	
Tilia grandifolia	Burger 1941	L – R	
Tilia japonica	Ohkura 1958	O – R	
Tilia parvifolia	Burger 1941	L – R	
Triplochiton scleroxylon	Scheiber 1973	Int.	Regular
Tristania conferta	Bootle 1983	Int.	
Trochodendron aralioides	Ohkura 1958	L	
Viburnum furcatum	Ohkura 1958	L	
Zelkowa serrata	Ohkura 1958	O	

[a] O = straight grained or less than 2°; L = LH spiral; R = RH spiral; W = weak; Int. = interlocked grain a common feature.

descriptions of *Quercus robur* and *Q. sessiliflora* (1947) and of *Fagus sylvatica* (1950a) showed that 70 to 80% of all trees in each of these species have maximum spiral angles less than 3°. Similarly, the study of *Fraxinus angustifolia* by Erdesi (1957) found spiral grain in only 17% of the logs examined, and the object of this study was to point out the *frequency* of spirality in a species where it had been commonly ignored. There are, of course, some species such as *Alnus rubra* (Kennedy and Elliott 1957) in which trees with straight grain are uncommon, and in which stem patterns more nearly correspond to the reversal of that in softwoods.

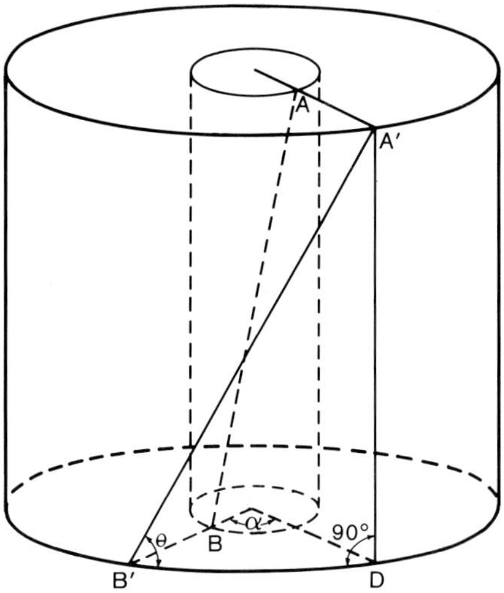

Fig. 1.2. Diagram showing how an initial spiral on a small stem at *AB* will increase with increasing diameter to *A′B′*. B′D/A′D = cot θ; B′D/B′O = α. Hence, B′O = (A′D/α); cot θ. Since A′D/α is constant, then cot θ is proportional to the radius B′O. (After Priestley 1945)

Looking at the situation as a whole, however, it is not difficult to understand the reaction of writers such as Priestley (1945) who concluded that spiral grain in hardwoods is fundamentally different from that in softwoods. Studying *Sambucus nigra, Syringa vulgaris,* and *Castanea sativa,* Priestley found that if the grain in the first year's growth followed an oblique course this was maintained through successive increments in the geometrical sense, so that grain angle increased simply as a consequence of the increasing girth of the stem (Fig. 1.2).

It is immediately apparent from the reversals of spiral direction, which do occur in hardwoods as in softwoods, that geometrical increase is not of universal application to hardwoods. Nevertheless, the concept raises two very interesting and quite fundamental questions.

The first concerns the application of geometry to the enlargement of woody stems which are, after all, produced by repeated, and in a sense independent, cell divisions. As the number of cells around a circumference increases by anticlinal divisions in the cambium to meet the requirements of growth in girth, what justification is there for assuming that succcssive cells produced by cambial initials will re-orientate geometrically to retain their relative positions? Priestley's measurements demonstrating that the cotangents of the grain angle show a linear relation to girth in a stem of *Sambucus nigra* certainly provide some support for the concept in this instance.

The second question is really the corollary of the first, relating to species that are predominantly straight grained or in which grain is restored to the axial direction after temporarily becoming oblique. Since minor twists and grain deviations must inevitably occur in the life of every tree, what is the nature of the process which restores straight grain (e.g., after branch stub occlusion) in the face of the geometric tendency to exaggerate all such deviations?

From Table 1.3 it can be deduced that in the majority of hardwood species spirality does not increase geometrically with increasing stem diameter and is therefore under some other type of influence. It may be that inability to reduce oblique grain arising by chance in the leading shoot is the *abnormal* characteristic of species such as *Sambucus nigra.* If this is so it deserves closer investigation to discover in what respect this species lacks the normal control over grain direction. It is of interest that Ohkura (1958) reported that *Sambucus sieboldiana* produces both LH and RH grain directions, apparently at random, though these did not exceed 2° in the ten specimens he examined.

For the present it need only be remarked that some trees appear to lack the ability to restore chance grain deviations to the axial or prevailing spiral direction. The frontispiece provides an extreme example of the consequences of this type of "accidental spiral" in a stem section of *Hebe* sp., a genus which, incidentally, has no wood rays.

1.1.2.3 A Three-Dimensional View of Spiral Grain

Whilst on the subject of the possibly diverse nature of different aspects of spiral grain, it is appropriate to mention experiments to study the movement of injected dyes along stems (Greenidge 1955, 1958, Vité 1958, 1959, 1967, Vité and Hendrickson 1958, Vité and Rudinsky 1959, Rudinsky and Vité 1959, Hendrickson and Vité 1960, Kozlowski 1961, Kozlowski and Winget 1963, Kozlowski et al. 1965, 1967). These experiments had various objectives, including studies of the transpirational water pathways within trees and the distribution of injected systemic insecticides. Previous work from 1901 onwards which led to these experiments was described by Vité (1959). The environmental and physiological significance of this body of work will be referred to in Chapters 4 and 7, but the patterns of water conduction that arise from spiral grain are of relevance here, and provide a particular insight into spirality as a three-dimensional system.

The method widely used is to inject dye (acid fuchsin is commonly used in a 0.3 to 1.0% aqueous or alcoholic solution) through a hole bored radially into the base of a stem, or by means of an injection knife. After a few days, distribution of dye within the tree is studied by sectioning the trunk transversely at regular intervals above the point of injection. In ring-porous hardwoods upward movement of water is largely confined to a thin cylinder of outer sapwood, but in diffuse-porous hardwoods and in softwoods a much greater proportion of sapwood is involved (Kozlowski et al. 1967).

The first point to be made is that dye usually follows the grain pattern, and the stained pattern obtained is therefore identical with the arrangement of the vascular conduits within a stem (Vité 1958, 1959). Vité (1967) referred to evidence that deviations can occur, depending on transpiration water deficits and pressure gradients, or in the presence of extreme spirality when unusually frequent bordered pits may encourage axial flow across the tracheids (Hartig 1895, Liese and Ammer 1962). But for moderate grain angles, in healthy stems, under normal conditions of transpiration, it can be assumed that the patterns of staining closely reflect the consequences of spiral grain (Kozlowski et al. 1967).

Stain patterns are said to indicate five types of water-conducting systems, as illustrated in Fig. 1.3. These are seen most clearly in those parts of the stem that lie within the crown. Where water movement passes through a considerable length

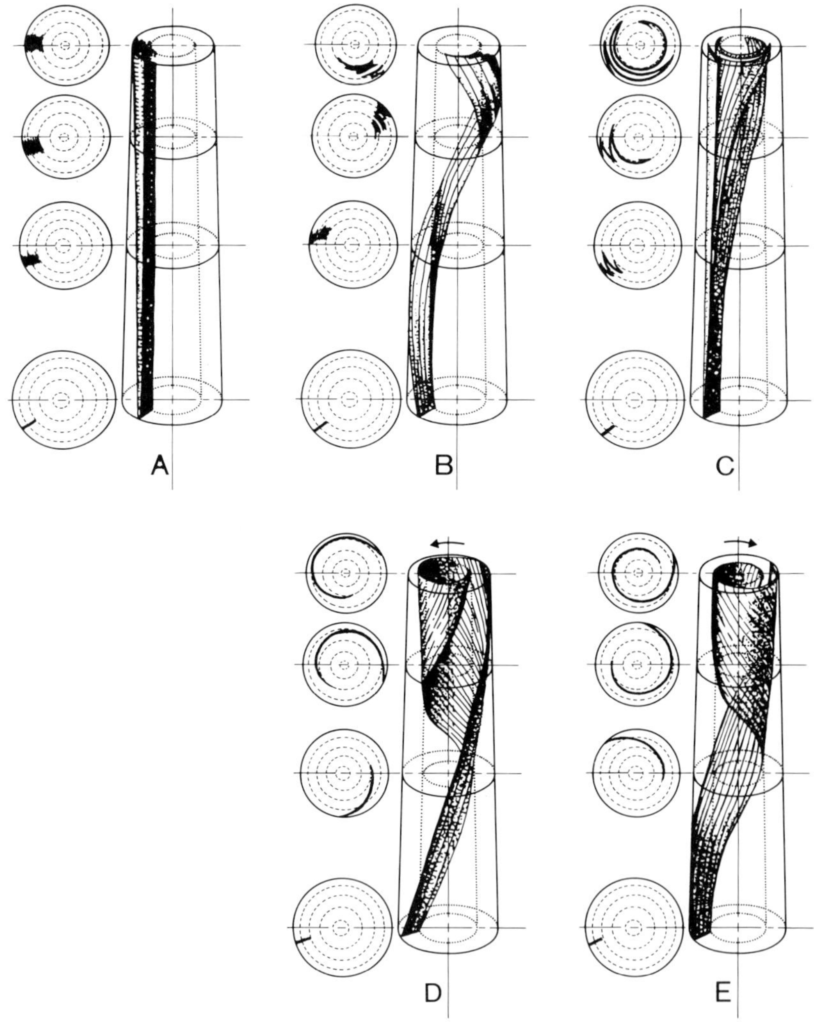

Fig. 1.3 A–E. Ascent of dye solution in stem sections varying in length from 1.05 to 4.20 m after injection radially at the base of the stem. **A** Sectorial straight; **B** sectorial winding; **C** interlocked; **D** spiral turning left; **E** spiral turning right. (After Vité 1967)

of lower bole, changes of fibre alignment and late reversals of spiral direction can obscure the basic system under consideration here. In other words, young trees and the crown region of actively growing middle-aged trees provide the clearest results.

Although some variations may be encountered within a single species, there is a tendency for dye patterns to be consistent both within a species and within closely related groups of species. However, Vité (1967) acknowledges that some species including *Taxus brevifolia, Pinus monophylla,* and *Pinus banksiana* have transpirational pathways that do not fall into any of the five basic systems.

The species groupings that are associated with these systems as set out by Vité (1967) are:

A. Sectorial straight ascent: *Thuja* and *Chamaecyparis*, in which dye ascends directly above the point of injection more or less parallel to the stem axis in all layers of the sapwood.
B. Sectorial winding ascent: *Pseudotsuga menziesii, Tsuga heterophylla,* and *T. mertensiana* [however, Kozlowski and Winget (1963) reported that *T. canadensis* has a more variable pattern]. In this system a stained sector winds around the tree as distance increases above the point of injection.
C. Interlocked ascent: *Sequoia, Taxodium, Libocedrus,* and *Juniperus*; also common in hardwoods (*Magnolia, Liquidambar,* and the large numbers of tropical hardwoods with interlocked grain). Vité noted that spiral angle may vary largely around the stem circumference, which confirms the findings of Knigge and Schulz (1959). The injected dye forms a zigzag pattern radially that extends its extremities as distance increases above the point of injection.
D. Spiral ascent turning left: In white pines, *Pinus lambertiana, P. monticola,* and *P. strobus*. When fully developed at a distance from the point of injection, the stained pattern as viewed from above a transverse section forms a (flat) spiral band that runs counterclockwise towards the centre of the stem.
E. Spiral ascent turning right: *Abies, Larix, Picea, Pinus ponderosa, P. taeda, P. resinosa,* and many other hard pines. The flat spiral band as viewed from above a transverse section runs clockwise from the periphery towards the centre.

It is worth emphasising that these are typically outer sapwood patterns for wood within the crown. Spirality in the innermost growth layers, heartwood (or not injected), is seldom indicated. Changes of spiral direction in old age, as in *Pseudotsuga menziesii* (Northcott 1957), will produce quite different patterns in the lower bole, changing from sectorial winding to right-turning semi-spiral (Vité and Hendrickson 1958, Rudinsky and Vité 1959).

The relationships between these three-dimensional systems and the stem patterns of spirality summarised in Tables 1.2 and 1.3 are self-evident in the sense that the three-dimensional patterns evolve from the direction of the initial spiral, and its rate and direction of change. In that they are also dependent on these relationships at all levels along the stem, they are by no means readily predictable from a single description of spiral angles for one transverse section. Even so, there are apparent anomalies between the two types of results. For example, *Larix kaempferi* is a species that has been extensively studied by conventional means, and all the results indicate an initial LH spiral with maximum values developing between the second and fifth growth layers from the pith, and with reversal to RH spiral usually occurring between the 27th and 35th annual growth layers (Nakagawa 1972, Ozawa 1972). The inference from dye injection experiments is that maximum LH grain angle is reached in the uninjected (heartwood) central region of the stem and that decreasing LH spirality in sapwood produces the "Spiral Ascent Turning Right" in the dye pattern as seen in a transverse section viewed from above. Note, however, that the direction that the dye follows along the stem is that of LH spiral grain, turning more slowly at the outside of the stem than towards the centre. In other words, direction of the *change* in grain angles is more important than actual spiral direction in determining the final spiral direction in dye patterns. Decreasing LH angles or increasing RH angles produce the same dye pattern (Vité 1958). Had the dye been injected into that portion of the stem above the level at which heartwood had developed, and in which LH spiral grain is building to its maximum before decreasing, different dye patterns might well have been produced. As it happens, dye patterns in this section of the stem cannot be identified unequivocally in any of the reports within this considerable body of work.

These considerations do not detract in any way from the significance of the three-dimensional systems within their own parameters. Vité (1967) argued that the grain patterns are genetically fixed and are specific to related genera or closely related species. Such individual variation as occurs he attributed to the environmental conditions that affect transpiration. From his observations he deduced that the general phenomenon of spiral grain should not be treated collectively. The five different systems of alignment in conducting tissues as revealed by patterns of injected dye should be accepted as being fundamental. It may be noted that this does not separate hardwoods and softwoods.

Vité did urge the need for further investigations of differences within the five systems. He recognised that species differ in the regularity with which patterns are developed within and between growth layers, and this might indicate further subdivisions within the original systems. In particular he recommended the three-dimensional approach to avoid "the conflicting interpretations in spiral grain research (which) stems from comparison of two-dimensional observations on unrelated species".

1.1.2.4 Pattern Down an Annual Increment

There is one further way of viewing changing spirality within a stem which has not received much attention in the literature, but which has particular relevance for defining those phenomena which should be included within the concept of spiral grain. This refers to the changing grain angles which develop from the wood formed within the leading shoot at the top of the annual increment right down the stem to its base. Species in which angles are consistent for wood of the same "age of formation" (i.e., for wood laid down by cambial cells of the same age) will have the same pattern of grain angles down a growth layer from top to bottom as that which develops radially from the pith out to the same growth layer at the base of the stem.

The typical LH to RH pattern of spirality in a softwood will illustrate this feature (Fig. 1.4). At the base of the stem wood will be straight grained next to the pith, with LH spirality developing towards a maximum at x rings from the pith, then decreasing to zero angle (straight grain) at y rings from the pith, and thereafter RH spiral grain increasing out to the cambium. Turning now to the outermost growth increment, primary and first-formed secondary xylem at the tip of the apical shoot are straight grained, but LH spiral grain develops lower down reaching a maximum value at x years' growth below the apex. Further down the

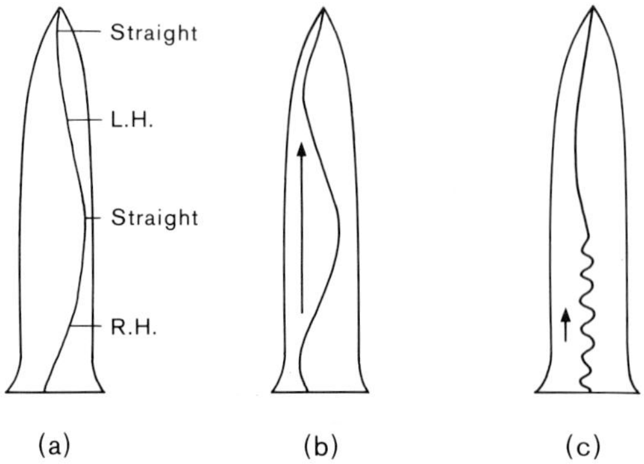

Fig. 1.4a–c. Grain direction along the outside of stems. **a** Typical softwood: from the top of the stem to the bottom grain direction describes three-quarters of a waveform. **b** Tree with interlocked grain: often more than one waveform, which moves a considerable distance up the stem in succeeding growth layers, so that grain angles at any given level change direction frequently. **c** Tree with wavy grain: often localised, closely compressed waves, which move slowly up the stem so that grain direction at any given level changes direction slowly

stem the grain angle will decrease to zero at y years' growth below the apex, and from there to the base of the stem there will be increasing RH spiral grain. A line following grain direction down the outside of the stem from top to bottom will describe three-quarters of a complete wave form (Burger 1941, Hendrickson and Vité 1960, Brazier 1965).

Similarly, trees that develop spiral grain (in either direction) increasing to a maximum and then reducing to straight grain will produce half a wave pattern from top to bottom of the outside of a mature stem. Trees in which grain angles increase from the centre of the stem to the outside may display only the beginning (perhaps a quarter) of a wave pattern. At all events, the occurrence of a "simple spiral" throughout the stem is most unusual in trees that produce significant spiral grain, e.g., with angles in excess of 2°. The pattern developed most commonly if the wood grain is followed from the top to the bottom of a stem is some part (usually between a quarter and three-quarters) of a wave-form.

1.2 Interlocked Grain

Interlocked grain is a more complex grain pattern, but one which has been widely accepted as being directly related to spiral grain because it has the appearance of reversals of spiral direction at more or less regular growth increments across a stem radius. The names by which it has been known at various times reflect this. In German it is called "Wechseldrehwuchs" (changing spiral growth) and Martley (1920) called it "Double Cross-Grain".

At times interlocked grain may arise to some extent randomly or by chance. This is because change of spiral angle across a stem seldom proceeds absolutely smoothly. When plotted graphically, angles in successive growth layers do not necessarily produce a straight line or a smoothed curve. It follows that in those species which change from one spiral direction to another, successive growth layers in the zone where the changeover occurs may fluctuate between LH and RH spirality for some years until the new predominant direction prevails.

Reversal of grain direction is not essential to produce interlocked grain (cf. Tree 3 in Fig. 1.5). When the degree of spirality fluctuates widely across a radius,

Fig. 1.5. Change of grain angle on the outside of three trees of *Liquidambar styraciflua*. (After Webb 1967)

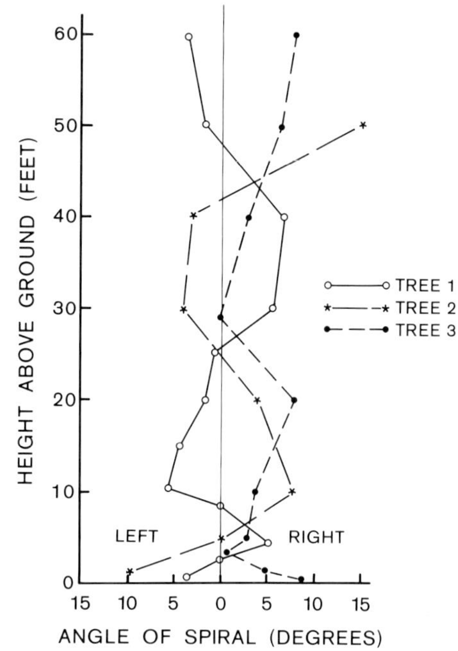

split samples will show interlocked grain even though all zones run in a constant (LH or RH) direction along the stem (Webb 1967).

Most of the interlocked grain reported in softwoods seems to have arisen from random variability rather than from definite and regular reversals of spiral direction (e.g., Harris 1965), though some trees undoubtedly show more erratic variations in grain angles than others. Ohkura (1958) indicated that 6 of the 33 Japanese softwoods that he studied had occasional interlocked grain (his "splitting curve" types S VI and Z VI) though extreme angles rarely exceeded 4°.

In general, interlocked grain in softwoods lacks the regularity and widely divergent angles of the extreme type of interlocked grain that is associated with many tropical hardwoods and with some genera of the temperate zone such as *Nyssa, Liquidambar, Ulmus,* and *Platanus.* Martley (1920) measured changes of grain angle in *Shorea robusta* as high as 32° LH to 32° RH in successive maxima over a radial distance of 37 mm, and in *Chloroxylon swietenia* from 40° LH to 40° RH. Harzmann (1965) measured a reversal from 27° LH to 41° RH in *Liquidambar formosana.* Hardwoods in which well-developed interlocked grain is a regular feature are included in Table 1.3.

On a smooth quarter-sawn (i.e., radial/longitudinal) wood surface interlocked grain produces the effect of dark and light longitudinal stripes depending on whether the grain is running downwards into the piece or emerging upwards out of the piece from the viewpoint of an observer at one end. It follows that when viewed from the opposite end the apparent shading of each stripe will be reversed.

Such is the regularity and closeness of these alternating stripes in species such as *Afrormosia elata, Entandrophragma cylindricum,* and *E. candollei* that they are much sought after for furniture manufacture in which this grain pattern can be used most attractively (Scheiber 1973). Closely related species can differ and Scheiber used the example of *E. angolense* which usually has relatively broad and less regular stripes than the other two species, though there are occasional trees in this species too with very close and extremely distinctive interlocked grain.

In all, Scheiber (1973) listed 34 species (or commercial species groupings) amongst the tropical hardwoods of commerce in which interlocked grain is a regular feature. This list, incidentally, does not include any Australian hardwoods amongst which interlocked grain is encountered as frequently as in any geographical grouping, in genera such as *Eucalyptus, Angophora, Tristania,* and *Flindersia* (Bootle 1983). Kadambi (1951) added two Indian hardwoods in which he said that interlocked grain is frequently encountered (*Mesua ferrea* and *Poeciloneuron indicum*) and one in which it is occasional (*Palaquium ellipticum*). Limaye (1954), on the other hand, reported that interlocked grain is of sporadic occurrence in *Mesua ferrea, Hardwickia binata, Michelia excelsa,* and *Anogeissus acuminata,* but is regularly encountered in *Shorea robusta* and *Cedrela toona.* This illustrates the point of considerable intraspecific variation and the varying experience of wood technologists in different areas.

Although the general appearance of interlocked grain gives an impression of alternating spirals, and although it has often been illustrated as such (e.g., by Scheiber 1973), the spirals are not constant along or around a growth increment (Martley 1920, Miller 1934). In fact, when grain angles are traced along a growth increment a sinuous pattern is followed (Fig. 1.4). The length of stem required to produce one complete sinuosity or waveform (e.g., from straight grain through LH grain angle, changing to RH grain angle and returning to straight grain) often varies along a stem, and the average wavelength also differs between species.

Webb (1967) showed that wavelength in *Liquidambar styraciflua* could vary from 3 to 12 m in a single tree, the fluctuations tending to lengthen from the base of the stem to the top (Fig. 1.5). Krawczyszyn (1972) examined the three species of *Entandrophragma* referred to above and found that the shortest waves on a tangential surface were 50 – 100 mm long and the largest 2 – 3.5 m.

Interlocked grain, like spiral grain, therefore involves changing grain angles of a wavelike nature down a growth increment, or, in the words of Krawczyszyn and Romberger (1980), some “manifestation of long-term rhythmic or cyclic processes in the cambium”.

The temporal rhythm which determines the frequency of grain reversals across a radius usually bears little relationship to the annual growth increments, but Détienne (1979) discovered that in *Daniella* species interlocked grain is of annual recurrence and, curiously, that in different species the development of grain reversal bears a different relationship to the beginning and end of the annual increment.

1.3 Wavy Grain

In interlocked grain the wavelength of grain sinuosity along a growth increment can vary considerably, but as long as the reversal of grain direction across suc-

Fig. 1.6. Radial/longitudinal surfaces of wood grain figures, illustrating the relation between near-transverse waves and near-vertical banding or stripe figure. Stem centre is to the *left* in all these examples. (Hejnowicz and Romberger 1979). 1 *Fraxinus excelsior*; 2 *Alnus glutinosa*; 3 *Betula pendula*; 4 *Platanus acerifolia*; 5 *Guibourtia ehie*; 6 *Entandrophragma cylindricum*

cessive growth increments is more apparent than longitudinal wave patterns on the tangential surface of growth increments, interlocked grain retains its essential characteristics. However, when the longitudinal wavelength decreases to the point where changes of grain direction along a growth increment attract attention the phenomenon is called wavy (curly) grain. Wavy grain is only rarely a regular feature within a species but tends rather to be occasional or intermittent.

Hejnowicz and Romberger (1979) used a series of wood samples to demonstrate the relationship between near-transverse waves on the radial/longitudinal surface of *Fraxinus excelsior* and near-vertical banding in *Entandrophragma cylindricum,* with examples of banding at intermediate angles in four other species (Fig. 1.6).

If the closely-packed wavy grain as found in *Fraxinus excelsior* is examined closely, waves on the tangential surface will be seen to have a very short wavelength, but a particular wave does not occupy precisely the same position across an entire radius. Instead, the wave pattern moves slightly up the stem through successive growth increments. Thus the radial stripes in this most intensive form of wavy grain form a small angle with the wood rays. In more open types of wavy grain, sometimes referred to as "rippled" grain, wavelengths on the tangential face of growth increments are longer and the waveform moves more rapidly up the stem per unit of radial increment than in *Fraxinus,* so that banding on a radial/longitudinal face forms broader stripes at a greater angle to the wood rays.

Eventually, as wavelength continues to increase, and rate of movement of the wave pattern up the stem increases across equivalent radial increments, a truly interlocked grain as in *Entandrophragma* is obtained. Yet even though the angle between the stripe figure and the wood rays appears to be 90° it is slightly less than that, and stripes do cross the annual increments if these are followed for a considerable distance. Figure 1.4 compares these different wave patterns.

Table 1.4. Relationship between wavelength of grain deviations along a growth increment and the velocity of their movement up the stem through successional growth increments. (Krawczyszyn 1972)

Grain type in wood	Species	Domain[a] height (mm)	Velocity of domain (mm mm^{-1} of radial increment)
Wavy	*Acer platanoides*	2–8	0.1
	Fraxinus excelsior		
Rippled	*Betula pendula*	10	1.0
	Picea abies		
Interlocked	*Platanus acerifolia*	80–280	6.5–12

[a] Strictly a "domain" refers to an area of cambium in which changes of grain direction leading to wave formation occur (cf. Chap. 5).

It was Krawczyszyn (1972) who postulated the direct relationship between wavelength and rate of movement of wave fronts up a stem, by studying the heights of zones in which changing grain angles are initiated (referred to as "domain height") and the rate at which the associated anatomical features move along the stem in successive growth layers (Table 1.4). The anatomical events that are the basis for these phenomena are the subject of Chapter 5. For the moment we are concerned only with the relationships between interlocked and wavy grain which have been so convincingly explained by Hejnowicz, Krawczyszyn, and their co-workers over the past 2 decades.

Waves in grain patterns have similar attributes to waves in other systems in that they have measurable length, amplitude, and period. It is now suggested (Hejnowicz and Romberger 1979) that more than one wave system may be present during wood formation, and that even more complex patterns than wavy grain (fiddle back) and interlocked grain (stripe) may result from superposition and interference phenomena that are features of waves in other physical systems. Grain patterns open to interpretation in this way include "blister", "roe", and other complex patterns. Baumer (1952) described many of the more complex wood figures, and also provided a glossary of terms in five languages.

If the waveform is accepted as a feature of certain types of spiral grain, and if it is accepted that this creates an affinity with interlocked grain, there can be no doubt that all manifestations of wavy grain are at least worthy of inclusion for study as having sufficient similarity to deserve explanation. This does not presuppose that all forms of spiral grain should be lumped together as identical phenomena but, by being able to differentiate between related forms, and by gaining some understanding of their various modes of formation and causation, we should be able to gain clearer insight into what spiral grain really is.

1.4 Other Growth Patterns

If such a wide range of grain patterns can be accepted as having possible affinities with spiral grain, it may be asked whether everything that is not straight grained should also be included, or, if not, what the grounds for rejection should be. To answer this we need to return to the starting point. We began with spiral grain, and then discovered that the spiral pattern in trees is seldom simple, but usually involves changing grain angles down a growth increment, which describe a wave pattern or partial wave pattern. Later, we shall see how this wave pattern derives from events in the cambium and that these events are not random but ordered. As a result, although grain angles change, they do so in a regular manner that is repeated (albeit with some inevitable biological variability) in other trees of the same or related species.

To express this as simply as possible, we are concerned with grain angles, following a non-random pattern created at the time of cell formation, and dependent on gradual and regular re-alignment of successional longitudinal wood elements. This may be a long way removed from simple spirals winding around a log, but it covers a range of patterns that share significant features related to their origins and development.

The final part of this chapter will look briefly at some of the other growth patterns that depart from the ideal of straight grain in straight trees but which do not fall within the definition given above.

1.4.1 Spiral Bole and Spiralled Branches

The term "spiral bole" has been used by Zobel (1965) to describe the situation in which the whole main stem of a tree has the form of a corkscrew, and which is seen in extreme form in the horticultural curiosity *Salix matsudama* cv. Tortulosa. This feature is occasionally found in certain conifers (Pemberton 1929), and Lasschuit (1951) described it as being the almost invariable condition in young stems of *Pinus merkusii* in northern Sumatra, in Indonesia. Though trees with spiral bole often contain spiral grain, as do straight-stemmed trees of the same species, there is no indication that the two are related in any way (Braun 1854), and Zobel (1965) suggested that they probably arise from different causes.

On the other hand, Canning (1915) found that the only way to tell which trees of *Pinus roxburghii* had RH spiral grain in the upper stem from any considerable distance was by observing the contorted form of the branches. Trees with LH spiral grain have a bulbous base and are therefore easily recognised even without reference to the branches which are also grossly contorted. All these spiralled branches contain severe spiral grain. In general, however, external evidence of spiral grain tends to be unreliable and examination of stem wood by one means or another is accepted as being necessary for the majority of species (e.g., Lowery 1967a).

1.4.2 Birdseye and Dimples

There are various grain patterns that originate from localised inactivation or reduction of cambial activity. The birdseye figure in *Acer saccharum* arises from inactivation of small areas of the cambium so that normal growth around them leaves the less active areas in shallow depressions (Jane 1956). It has been suggested that a parasitic fungus attacking the cambium may be responsible.

In some pines occlusion failure above the embedded region of the short shoots (needle traces) after stem needles have been shed, leaves behind a series of depressions in the cambium to the extent that the wood so affected may contain patches of included bark. In *Pinus radiata* this is sometimes called birdseye, but is obviously different in appearance and origin from that in *Acer,* and although the pattern is spiralled (following stem phyllotaxis) it does not depend on regular and ordered re-orientation of the stem tracheids as a whole.

The dimpled grain in *Pinus contorta* arises in an entirely different way, owing its origin to resin blisters produced in the phloem which retard the activity of the cambium by pressing against it (Jane 1956).

1.4.3 Curly Birch

The northern coniferous zone in Europe contains very few tree species which possess wood with ornamental properties. This is probably why so much attention has been paid to the various unusual growth patterns that occur in *Betula* spp. Etholén and Huuri (1982) have compiled a bibliography containing 442 titles, of which 162 are in Russian, dealing only with the various types of decorative grain patterns in *Betula.* Only five of these papers are included in the bibliography to this book to provide some representative coverage, because it is contended that grain patterns in *Betula* (normally a very straight-grained genus) do not fall into the category of spiral grain.

Ruden (1954) described three grain patterns in *Betula pendula* which he called speckled, maser, and burl figures, whereas Vaclav (1969) referred to birdseye, flamey, and curly figures. The presence of these grain patterns can usually be recognised from external features such as curly, ridged, or scabrous bark, clusters of undeveloped buds, and unusual stem form (Newall and Gardiner 1963, Vaclav 1969). Curly and flamey birch can be found only in *Betula pendula* whereas birdseye can occur in several species including *B. pubescens* and *B. kylowii.* Not only are these grain patterns quite strongly heritable (Johnsson 1950, Heikinheimo 1951, Persson 1954) but the various forms of *Betula* spp. producing them have been given taxonomic recognition.

Wood of curly birch is described as containing V formations with the points turned inwards. These contain anatomical irregularities and multi-layered rays following an irregular but "nearly spiral" course tangentially. Birdseye patterns contain numerous rays and fewer vessels, but the surrounding wood is nearly normal. Flamey birch has shorter fibres with thicker walls forming an irregular pattern, and wood rays are more numerous (Vaclav 1969). None of these features depends on the regular and ordered re-orientation of longitudinal stem elements that produce spiral, interlocked, or wavy grain.

1.4.4 A Collection of Oddities

There can hardly be a single woody species which, if studied extensively, would fail to yield the occasional abnormal, decorative, or bizarre grain pattern. These may derive from crotches, burls, knots, resin pockets, and kino veins, or from ingrown phloem, bark, and stem cones – all features whose origins are fairly obvious – or they may be of indeterminate origin such as "gnarliness" in *Pinus sylvestris* (Ahola 1952) or mottled wood in *Agathis australis* to pick an example from the Southern Hemisphere.

None of these grain patterns relates readily or directly to the essential features of spiral grain, and the nature of "other grain patterns" has been pursued in some detail only to emphasise those features exclusively relating to spirality that will provide the basis for further study.

That is not to say that if we knew more about the anatomy, physiology, or heritability of other grain patterns it could not assist in our understanding of spiral grain. Quite simply, no relationship has yet been demonstrated and there are no prima facie reasons for pursuing these issues further.

1.5 Conclusions

1. Spiral grain in its simplest form appears on the outside of affected stem sections, in which the longitudinal wood elements are not strictly parallel to the stem axis but run around it in a spiral. This spiral is seldom consistent from the centre to the outside of a stem. The spiral pattern may also change along a growth increment, and often traces partial or multiple waveforms when followed along an entire stem from bottom to top.
2. There are grounds for including interlocked and wavy grain in any study of spiral grain, because they share these features of longitudinal waveforms dependent on gradual and regular re-alignment of longitudinal elements across successive growth increments, and we shall consider the possibility of interactions between waveforms.
3. For completeness, straight grain should not be ignored because it represents the maintenance of zero spiral.
4. None of the other multifarious grain patterns that have been described meet these criteria and will therefore receive no further attention.

Chapter 2

Measuring Grain Angle

2.1 Detection of Grain Angle

In this chapter the term grain angle will be used from time to time to refer to deviations other than those due to spiral grain or its related phenomena of interlocked and wavy grain. Sloping grain in sawn timber, commonly called cross grain (Anderson et al. 1945), can arise in many ways. Grain deviations around knots and other defects are an obvious source, as are all types of stem kinks including spiral bole and nodal or other stem swellings. In addition, grain will run across sawn timber if the cutting pattern is not parallel to the growth layers. Spiral grain itself affects grain angles only in the tangential plane relative to the stem axis. For practical purposes it may be of little consequence how grain deviations arise, but it is often important to be able to measure cross grain as the result of any or all of these sources (Boas 1933, Anderson et al. 1945, Koehler 1955).

The system of measurement and degree of precision most appropriate to a particular study will depend on several features including (1) the pattern and intensity of spiral development within a species, (2) the effect of grain angle on physical and mechanical properties, and (3) the importance of both these aspects for the proposed or potential uses. Two examples will illustrate how different approaches are appropriate to different studies. In a structural application such as the chord of a roof truss, maximum grain angle within the region of stress may be of prime importance: in effect, a single measurement provides the essential information on grain angle. When evaluating a tree to be used in a breeding programme, however, it may be necessary to make hundreds of measurements in order to assess the total pattern of spiral development within it, and complex analyses will be required to assess the significance of such patterns for potential end uses.

Before cross grain can be measured it must be detected, however, and even in timbers with distinct annual growth layers grain direction may be difficult to detect on any surface that is not strictly radial. For example, in the writer's experience the most frequent queries concerning spiral grain that came into his laboratories were broken parts of such critical members as ladder stiles, truss components, and frames of light aircraft. These were commonly made from close-grained softwoods with distinctive annual growth layers (e.g., *Pseudotsuga menziesii* and *Picea sitchensis*) and had fractured diagonally under stress as a result of undetected cross grain. The linear annual growth layers on quarter-sawn faces, and the regular pattern they presented on flat-sawn faces gave no indication that the tracheids ran other than strictly longitudinally throughout these pieces.

Numerous methods have been devised to detect the true orientation of wood grain. Some are of general application, whereas others are restricted to the circumstances for which they were developed. These will be described in Section

2.1.3, but first we will consider external visual evidence of spiral grain that can be identified in the field without resorting to equipment other than the human eye.

2.1.1 Detection in Trees and Round Produce

When patterns of spirality develop in a very regular way, ring by ring from pith to bark, the spiral pattern down the outside of stem may give some indication of grain angles in inner growth layers. Thus grain angles, measured on the outside of each annual height increment from the top of a stem downwards, will provide much the same sequence as grain angles measured in each annual growth layer from the pith outwards across a radius. How far the radial sequence develops will, of course, depend on the number of growth layers at any given level. The value of this relationship will be discussed more fully in Section 2.2.1.

Very regular spiral patterns are more likely to be found in young plantation-grown trees than in older trees in natural stands, though even in old trees observations of exterior grain angles could prove useful, depending on the growth characteristics of the species. For example, straight grain on the outside of a mature pine tree would reveal nothing about the intensity or extent of the spiral grain that is characteristically developed in youth. On the other hand, many members of the Podocarpaceae tend to be straight-grained in youth, but spiral grain is an occasional feature of senescence. A mature podocarp with straight grain on the outside of the stem could therefore be assumed to be straight-grained throughout, but other observations would be required before spirality in the pine tree could be assessed.

When detailed information is required about the full range and intensity of spiral grain in a tree, decisions will have to be made about the damage that may be done in obtaining that information. Many of the methods described later in Section 2.2.2 have been developed to obtain as much information as possible whilst minimising damage to valuable timber. Unfortunately all methods are time-consuming and demanding. For this reason no external visual evidence of grain patterns should be ignored. Even a rough preliminary screening can save much work in studies where hundreds of trees have to be examined for comparative studies of environmental effects, tree breeding, and genetics.

2.1.1.1 Evidence from Radial Splits and Scars

Spiral grain in round produce and in dead trees without bark usually first becomes apparent through the development of drying checks. Other features such as large wood rays, vessels, and resin canals may also be useful indicators. In standing trees, scars from wind damage as well as drought and frost cracks (Krahl-Urban 1961) and lightning scars (Champion 1924) may provide evidence of the alignment of underlying wood elements.

2.1.1.2 Evidence from Shape of Stem and Branches

Many stems develop ridges that arise about the insertion of roots and branches. In spirally-grained trees these ridges may curve around the stem (König 1947, Krahl-Urban 1953 b, Mayer-Wegelin 1956) and the overall effect is as if the entire tree had been twisted about its axis. Reference to "twisted" trees and fibres so common from the turn of the century to the 1950s was a very natural, if rather misleading, response to this phenomenon (e.g., Cheney 1896, Bradley 1918, Champion 1924, Jacobs 1935, Kadambi and Dabral 1955).

In *Pinus roxburghii* Canning (1915) found that trees which developed a very bulbous base always had extreme LH spiral grain in that region. Twisted and contorted branches were found to be a feature of trees with both LH and RH spirality, though this too was more marked in trees with LH than those with RH spiral grain.

Krahl-Urban (1953 a, b, 1960, 1961, 1967) also described how twisted branches are a feature of trees with spiral grain in *Fagus sylvatica* and possibly, though less clearly, of *Quercus robur* and *Q. sessiliflora.* Moreover, he claimed that twisted branches show up in young trees well before spiral grain has developed significantly in the stem, and their presence can therefore be used in the decision to remove trees during thinning operations.

2.1.1.3 Evidence from Bark Patterns

Bark fissures and patterns of bark exfoliation have been related to the underlying grain directions in wood by many writers (e.g., Jones 1931, Wentworth 1931). No doubt intimate knowledge of a particular species enables the expert to detect signs that would not be perceived by the layman. McCarthy and Hoyle (1918) claimed that "spiral can be detected in the standing tree (of *Picea sitchensis*) by a trained expert without the removal of bark". Others have doubted the reliability of bark patterns in *Picea,* and König (1947) found that in *Picea abies* the process was usually difficult and highly unreliable.

One obvious difficulty is that the orientation of bark elements originating in the phloem provides a mirror image of orientation within the xylem. Outer phloem was laid down at the same time as the inner xylem. Thus the patches left by bark exfoliation of *Dacrydium cupressinum* in Fig. 2.1 serve to emphasise the pattern of stem ridges and furrows, but their orientation may be related to that of wood well inside the current year's growth. Canning (1915) explained how very slow bark exfoliation in *Pinus roxburghii* produces a considerable time lag, amounting to 60 years or more, between the external appearance of the bark and grain orientation in the underlying wood.

This type of time lag probably accounts for differences of opinion about the usefulness of studying furrowed bark in species like *Quercus robur.* Although the direction of bark furrows may generally indicate spiral grain, when present (Knigge and Schulz 1959, Birot et al. 1980), Mayer-Wegelin (1956) illustrated a frost crack originating in the outer growth layers of an old oak which was clearly running at a considerable angle to the bark furrows.

Fig. 2.1. Spiral ridges and patterns of bark exfoliation in *Dacrydium cupressinum*. (Photo Dr. C. L. O'Loughlin)

Pillow (1955) used a carpenter's draw shave to remove outer bark in order to look for wavy patterns in the inner bark. In this way he was able to identify the rare trees of *Juglans nigra* that had the curly grain so desirable for decorative furniture and panels.

An interesting point was raised by Newman (1956) with regard to the geometrical increase in spirality consequent upon circumferential growth, as postulated by Priestley (1945). Newman questioned whether this concept is necessarily applicable to a labile system such as the process of wood formation by the cambium. On the other hand, he pointed out that the older tissues of the bark are not labile. Growth has ceased in them, and they are moved bodily outward with increase in radius of the trunk. Under these circumstances increase in spiral angle would seem to be inevitable, and one would therefore expect the inclination of

Table 2.1. Spiral angles in wood and bark of *Cupressus* spp.

Species	Bark spirality (RH)	Height in stem (m)	Wood spirality (°) at rings from the pith							
			5	10	15	20	25	30	35	40
C. macrocarpa	Severe	0	1.1 R	0.2 L	1.5 R	2.1 L	1.6 R	0.0	0.6 R	0.2 L
		5	–	0.4 R	0.4 L	2.5 L	4.5 R	5.4 R		
		10	1.8 R	1.9 L	0.0	1.6 L	3.4 R			
	Severe	0	2.5 L	0.4 L	1.9 R	1.1 R	2.8 R	0.0	1.8 R	4.5 R
		5	0.9 L	2.9 L	2.9 L	0.8 L	1.2 R	3.1 R	3.2 R	
		10	3.6 L	4.1 L	1.9 L	3.0 L	1.0 R	2.4 R		
	Nil	0	1.5 R	1.6 L	1.5 L	0.8 L	2.0 L	1.9 R	2.0 R	4.4 R
		5	–	5.2 R	2.4 R	2.0 R	1.6 R	0.0	0.2 L	
		10	1.9 R	2.0 L	0.8 R	1.5 L	2.1 L	2.5 L		
C. lusitanica	Severe	1.4	2.8 R	1.0 R	0.5 R	2.2 R				
		5	8.5 R	8.0 R	8.0 R					
	Severe	1.4	5.2 R	0.5 L	3.2 L	7.5 L				
		5	3.5 R	1.5 R	2.0 R					
	Nil	1.4	0.2 L	1.8 L	0.2 L	1.2 L				
		5	0.0	0.5 L	2.0 L					
	Nil	1.4	0.0	1.5 L	3.2 L	2.0 L				
		5	1.2 R	4.0 L	5.0 L					

grain observable in bark to show a much greater angle than that in wood formed contemporaneously with it.

No measurements appear to have been made to check this relationship. It may account for a phenomenon reported by M. H. Bannister (personal communication) in which spiral bark patterns are found in *Cupressus* logs that have "straight-grained" wood. That is, a very small degree of spirality in the centre of the log may become exaggerated in bark formed at the same time. On the other hand, the orientation of areas of phellogen may not always be aligned longitudinally, and this too could lead to spiral patterns in bark. Both of these possibilities deserve further study.

I am indebted to Mr. G. D. Young (New Zealand Forest Research Institute) for providing me with previously unpublished measurements of spirality in wood and bark of *Cupressus macrocarpa* and *C. lusitanica* (Table 2.1). Two trees of each species were selected with markedly spiral patterns in their bark (more than 10° RH spiral) and one (*C. macrocarpa*) or two (*C. lusitanica*) trees with axial bark striations. Grain angles in the wood of *C. macrocarpa* were measured at every fifth growth layer from the pith outwards at 0, 5 and 10 m above ground level. In all three trees, irrespectively of bark pattern, grain angles in the wood tended to be small, and were equally to the left or to the right with no distinctive pattern. In *C. lusitanica* measurements were made on discs cut from 1.4 and 5.0 m. Although grain angles were generally larger in the trees with spiral bark than in the one with straight bark, one tree with RH spiral in the bark had changed to LH spiral in outer wood after initial RH spirality. In neither species did the bark give any clear indication of grain angles in the wood, though it could be argued that in *C. lusitanica* bark angles were more closely related to initial spirality in the wood.

2.1.2 Detection in Seedlings

Ideally, early detection at the seedling stage of trees destined to produce severe spiral grain would enable them to be removed from the nursery bed so that the forest would produce nothing but straight-grained trees.

Unfortunately, many tree species that are known to produce spiral grain during their lives have been recorded as producing straight-grained wood in the first year's growth (often in the sense of the first annual growth layer from the pith at *all* stages of growth), e.g., Champion (1924) and Misra (1939) for *Pinus roxburghii*; Jacobs (1935) and Harris (1965) for *P. radiata*; Mikami and Nagasaka (1975) and Kawaguchi et al. (1982) for *Larix kaempferi*.

Priestley (1945) suggested that all softwoods start off with straight grain. However, there are many reports of early development of spirality in both softwoods and hardwoods. Northcott (1957) described both LH and RH spiral grain in first-year seedlings of *Pseudotsuga menziesii, Alnus rubra,* and *Populus nigra.* Toumey and Korstian (1942) referred to Toumey's experiments of 1901 in which spiral grain occurred in 3.5% of first-year seedlings of *Pinus jeffreyi* derived from trees with spiral grain. Twist could even be recognised in the hypocotyls of these seedlings. Jablanczy (1969) also described twist in the hypocotyls of *Picea abies,* as did Kadambi and Dabral (1955) in seedlings of *Pinus roxburghii* derived from parent trees with severe spiral grain, although Champion (1924) had not regarded the incidence of twisted hypocotyls in the species as being significant. Northcott (1957) used seedling phyllotaxy to detect spirality, by measuring displacement of sequences that would normally be axial.

Not surprisingly, there are few long-term studies that have followed events through from seedlings to mature trees. The classical studies of Champion (1924, 1927a, 1930, 1931) are still an outstanding example in that they continued from seedlings (1- to 5-years-old) through to saplings aged 14.

In any case, initial development of spirality is only one part of the story. For mature trees, the total impact of spiral grain on wood use will depend on rate and duration of development from the pith outwards, maximum spirality in the initial direction, time taken to reach the maximum value, and possibly also time taken to change direction and maximum value in the new direction.

All these features appear to vary pretty much independently in all the species that have been studied so far, and therefore trends in seedlings can provide only part of the picture. However, if any seedling characteristic can be used to exclude or to reduce the frequency of the more extreme forms of spiral grain, this possibility should certainly be pursued with vigour. At the very least it can be said that a high proportion of "twisted" seedlings in species such as *Pinus roxburghii* are a sure indication that a poor seed source has been used. It would be better to destroy such seedlings before planting out, rather than commit "avoidable mistakes with which forestry will be saddled for a whole tree generation" (Champion 1945).

2.1.3 Detection in Wood

The processes used for detecting grain direction in wood need not be applied exclusively to sawn timber and debarked round produce, or used as a tool for timber research. Many of the methods that will be described have been also used on standing trees and seedlings, though most require some degree of mutilation such as bark removal in order to detect patterns in wood.

2.1.3.1 Natural Features

Natural features that can be used to follow grain direction have already been referred to and include orientation of vessels, resin canals, and wood rays. Some rather ingenious methods have been devised to follow the orientation of large vessels in ring-porous hardwoods. Knigge and Schulz (1959) inserted fine filaments into the vessels of *Quercus petraea* to register their orientation, and Tiemann (1941) was able to measure vessel direction in a board by forcing a soap solution into the vessels on one side and observing the point at which it bubbled out. Though they both clearly made their respective points, these methods have not been widely adopted.

Natural discolourations in wood such as resin streaks, stains, and kino veins may also follow or emphasise longitudinal wood elements.

2.1.3.2 Ink Test

Similar effects to natural discolourations can be produced artificially by using dyes such as safranin in alcohol, or any ink that has good flow characteristics (but not black drawing ink). The coloured solution, applied as a transverse line drawn across a tangential wood surface, spreads rapidly up and down the fibres producing a series of fine lines in the grain direction.

2.1.3.3 Dye Injection

Taking the ink test one stage further, dyes injected into the sapwood have been used to detect grain patterns in live seedlings (Northcott 1957) and standing trees (Vité 1958, 1959, 1967, Vité and Hendrickson 1958, Rudinsky and Vité 1959, Kozlowski and Winget 1963, Kozlowski et al. 1967). Slope of grain in the outer growth layer of wood can be detected from small windows judiciously cut in the bark above the point of injection to minimise damage, but assessment of the total pattern of dye movement through the sapwood naturally requires trees to be felled and dissected as described in Section 1.1.2.3.

2.1.3.4 Radioactive Tracers

Radioactive tracers ^{42}K (Keller et al. 1974, Arbex et al. 1978, Birot et al. 1979, 1980, Teissier du Cros et al. 1980) and also ^{188}Re (Arbez et al. 1978), have been used to enable grain angles to be detected without bark removal. The radioactive tracer is injected just below the cambium and is carried by the ascending sap in the direction of the grain. Its presence a metre or so above the point of injection is detected by applying a photographic film to the bark, and grain angle in the outer wood can be measured within 1° or 2°.

2.1.3.5 Scribe Test

The method that has been most widely used on a variety of materials including seedlings (Northcott 1957, 1958), trees, sawn timber, round produce, and veneers, involves forcing a sharp point a small distance into the wood surface so that when it is drawn approximately along the grain the true grain direction is indicated by a scratch mark. Koehler (1955) described an instrument that has been widely adopted and this is illustrated in Fig. 2.2. It comprises a steel bar that is freely pivoted in a wooden handle at one end and which has a gramophone needle protruding from the other. This "slope of grain detector" has been variously modified (e.g., Kano and Saito 1965) in efforts to make the instrument less subjective in application, to ensure that the point follows the grain direction as faithfully as possible, and to control insertion pressure.

However, it is probably fair to say that the simplest instrument will give satisfactory results and can be used with great accuracy as long as its limitations are understood and proper precautions are taken. Strongly defined latewood boundaries should be avoided because the needle may be diverted through softer earlywood. In all applications it is advantageous to make an initial trial so that the needle can be pulled as nearly as possible in the true grain direction. Thereafter it is good practice to make three or four marks to ensure that the grain runs parallel over the area to be tested.

Pelle (1967) described a knife having an extended point to the blade that can be used to detect grain direction in thin-barked trees. The point is dragged down the underlying wood and the blade cuts through the bark. When used on young *Pinus radiata* up to 200 mm in diameter the slope of the line cut through the bark

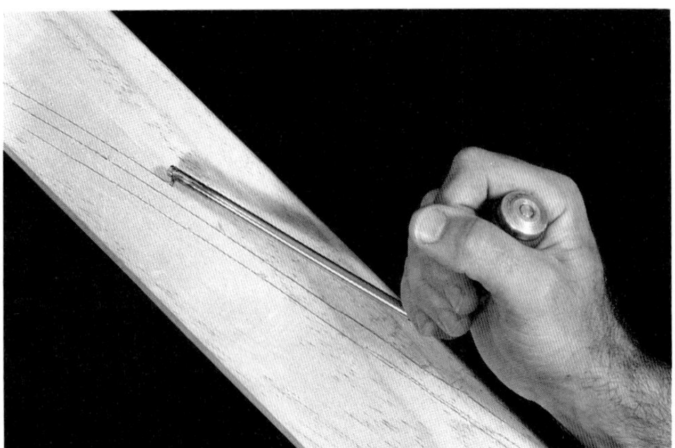

Fig. 2.2. Slope of grain detector. (Photo H. G. Heimcke-Hemming)

was said to be approximately 1° greater than the angle of the grain at the wood surface.

2.1.3.6 Pick Test

The pick test involves pulling at a small bundle of fibres to produce a line of separation along the grain. For large cants in the rough-sawn condition, surface splinters can be pulled free, or a small splint raised with a fine chisel or sharp screwdriver. However, when this test is used for research purposes, a fine pair of tweezers such as those used by watchmakers, and a wood sample with a moisture content above fibre saturation, can achieve considerable accuracy by removing as few as 30 fibres (Nicholls 1971 a). Nicholls recommended this as the preferred technique for studying small specimens such as might be removed non-destructively from standing trees, and he would restrict the scribe test for use on large specimens such as logs, boards, and veneers.

2.1.3.7 Splitting

Grain direction as revealed by splitting a log has been used by many workers. A wedge-shaped blade is applied to the end grain across a diameter and, after a split has been initiated, the blade is twisted to complete separation of the two halves of the log. The major study by Ohkura (1958) was based almost exclusively on split logs of a very wide range of species. Ohkura recognised six standard types of splitting curve each of which might be generated with the initial deviation to the right or to the left, and he described a further six "miscellaneous types" which represented special circumstances (Sect. 2.3.8).

When this method is applied to a log with interlocked grain, splitting is of course very difficult, though the results are most graphic (Martley 1920, Ohkura 1958, Scheiber 1973, Détienne 1979, Miyaki et al. 1983).

2.1.3.8 Sonic and Capacitance Properties

The longitudinal components of wood structure are responsible for differences in many properties when measured with the grain or across it. Various attempts have been made to use such properties to determine grain direction.

Foulger (1969) used a sonic pole-testing device, Pol-Tek, to determine the direction in which sound pulses travelled at maximum speed (i.e., with the grain). Observations made by inserting probes through the bark of standing trees indicated that the method could be used to determine slope of grain in the outer wood layers, but only on within 5°, and occasionally readings were more erratic than that.

McLauchlan et al. (1973) constructed a non-contact sensor that used the dielectric constant of wood for continuous inspection of grain direction in sawn timber. When a material is placed between two capacitor plates and subjected to an electrostatic field, dipoles within the material have a strong tendency to align themselves with the field. The dielectric constant of the material is an index of its ability to store electrical energy in this way and can be measured by the capacitance between the plates. The dielectric constant of wood is 1.5 times greater with the grain than across the grain, so that by detecting the field orientation that corresponds to maximum dielectric constant, grain angle can be measured, and this is independent of moisture content, wood density, or species. Application and use of the dielectric sensor have been discussed by Samson (1984), and James et al. (1985) described a microwave method for measuring moisture content, density, and grain angle of wood.

2.1.3.9 Microscopic Observation

The term "wood grain" refers to the alignment of longitudinal elements such as tracheids, fibres, and vessels. In this sense the ultimate test for grain direction is the actual examination of these elements. In conifers without longitudinal resin canals, and in many hardwoods, this can be achieved only by magnification. A good hand lens may be sufficient, but a low power microscope will usually provide greater accuracy.

It will often be necessary to examine successive tangential surfaces across the diameter of a stem in order to follow changes in spiral angle, so that the use of incident lighting will save the necessity for section cutting. As long as a reference direction having a known relationship to the stem axis can be defined for all samples, it is a relatively simple matter to align some ocular reference line with the longitudinal wood elements. Small-scale deviations such as those due to leaf traces in young stems may pose some problems but the average grain direction can probably be determined as accurately with a low power microscope as by any other means. Specimens should, of course, always be observed at their outer surfaces, otherwise the grain orientation becomes reversed.

2.2 Sampling for Grain Angle

Sampling for spiral grain implies measuring grain angles at a number of places in a stem or piece of timber sufficient to be able to predict grain patterns for the areas of concern within acceptable limits of error. Sampling usually assumes that grain angles follow some sort of pattern within stems. Obviously, if grain angles vary in a totally random fashion only generalised statistics can be obtained such as mean angle and some measure of variability. As we have seen, however, spiral patterns generally do develop in a way that is characteristic for the species so that sampling to predict spiral values becomes a possibility. Good sampling strategy requires clear definition of the characteristics that are to be quantified. Samples appropriate for evaluating spiral grain in a population will be different from those required to establish indices for individual trees, for example. Unfortunately the literature contains many examples of sampling for spiral grain without any clear indicaton of objectives or even that the probable limits of error have been calculated or indeed considered.

For some studies the sampling process may have to commence with tree selection. Whyte et al. (1980) indicated how this might be organised for a study ab initio with specific objectives. In young plantations of *Pinus caribaea* growing in Fiji, six age classes were selected for examination in the main forest areas so as to provide a broad coverage of the total resource. Plots representing age classes were randomly located within three forests with the probability proportional to their area, and a transect across each plot was set out to cover as much topographic variation as possible. Then at each of four systematically spaced sampling points per transect, the nearest and fifth nearest trees were selected to be measured, making a total of 144 trees, 8 from each of 18 plots.

In other types of studies, trees to be examined will be determined by other criteria, such as the requirements of tree-breeding programmes or progeny trials. However, there still remains the question of selecting sampling points within stems. Nicholls (1971 a) suggested that the mid-point of an internode or a position slightly above it is best, as this avoids grain deviations due to knots. Then grain angles should ideally be measured across two diameters at right angles, or, failing that, across one complete diameter commencing at a point where grain angle under the bark is at a maximum. This can be observed through windows cut in the bark. Within growth layers, Nicholls suggested that the last-formed latewood is the most suitable sampling point, being readily located, comparable in all growth layers, and usually representative of each growth layer as a whole.

The frequency of measuring points across a diameter, and the numbers of diametric samples taken from each stem will depend on the type of study, and whether destructive sampling is acceptable or whether the tree is to be injured as little as possible.

2.2.1 Destructive Sampling

Studies designed to investigate the basic patterns of spiral grain for a species will inevitably require very detailed measurements at numerous points along stems. In young trees, measurements at every annual growth layer across diameters taken from every annual height increment may be justified (Harris 1965). Nakagawa (1972) measured grain angles at every annual growth layer across diametric strips cut at 2-m intervals along *Larix kaempferi* trees aged 49 years. This species is regarded as having rather regular development of spirality within individual stems so that patterns developed from the pith outwards show considerable similarities at different heights in the tree (Fig. 2.3). The coefficient of correlation between the grain angle at breast height and that in comparable growth layers up the stem was statistically significant (r approximately 0.5). However, the implication that only 25% of the variation within the stem as a whole can be predicted from breast height samples indicates the variability likely to be encountered even in species with definite patterns of spirality.

Fielding (1967) found that grain angles in *Pinus radiata* also tended to follow a pattern characteristic of each tree, and that deviations from these patterns (such as major differences around a circumference) were rare. However, he too found it difficult to obtain consistently close correlations between grain angles measured

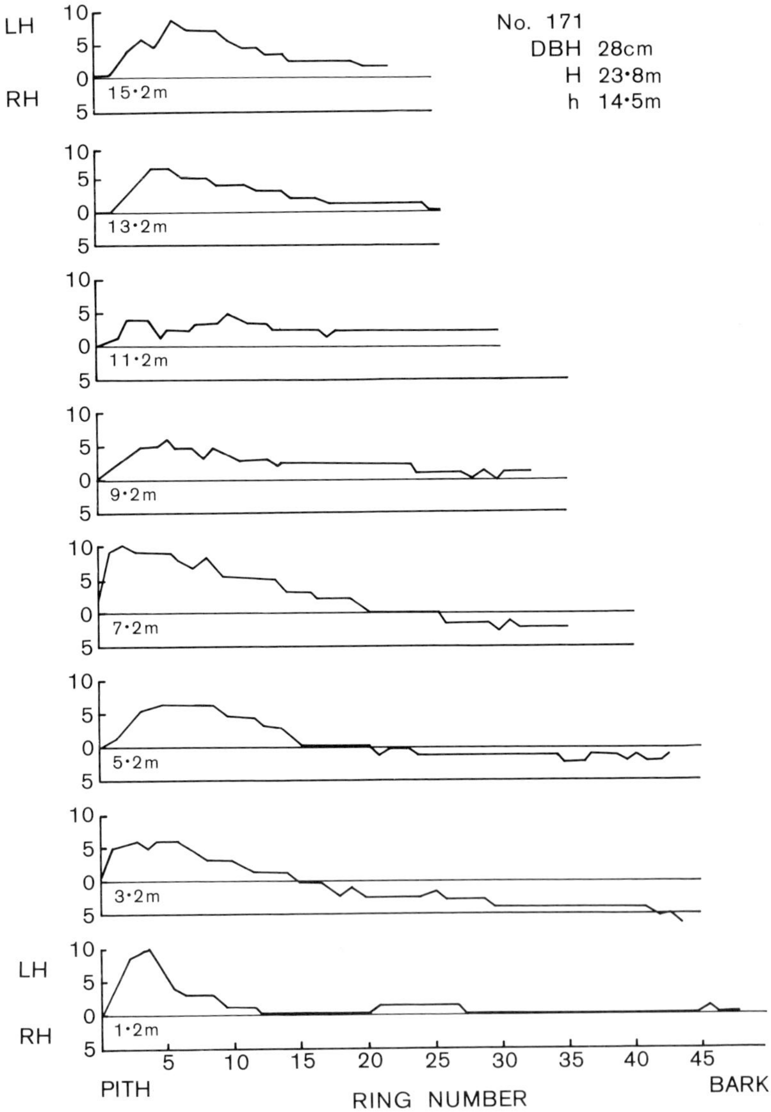

Fig. 2.3. Spiral grain at different heights in a stem of *Larix kaempferi.* (After Nakagawa 1972)

at surface sampling points and other (internal) measurements. He compared grain angles under the bark at ten points along each stem with the mean values for corresponding growth layers measured on diametric strips cut from 3-m intervals. Five of ten correlations were significant at $p<0.01$ and one at $p<0.05$ but the other four showed no significant correlation. Nevertheless, Fielding concluded that, although the angle of the spiral at a given number of annual rings from the pith varied considerably in some of the trees, measurement of spiral angle at a number of heights on the surface of the trunk gave a useful indication of the

Fig. 2.4. Weathered stump of a *Eucalyptus* sp. showing widely divergent grain angles above the insertion of roots, and considerable circumferential variation well above the region of basal swelling. (Photo H. G. Hiemcke-Hemming)

spiral angles within and could be used to discriminate between trees with severe spiral grain and those with straight grain.

Paterson (1967) also used measurements of grain angles under the bark to assess spiral grain in *Pinus patula, P. radiata,* and *Cupressus* spp. grown in East Africa. Windows were cut in the bark in the four cardinal directions and at 3-m intervals along the stem up to 11 m, the log length said to be of practical interest to East African forestry. For 26 trees of *Pinus patula* Paterson calculated that 20 observations would be required to estimate mean external angle within ±2° for $p<0.05$. He concluded "In 16 readings up a bole differences as much as 12° can occur between readings. Let this be a warning to those who think they can safely sample at one position, perhaps even with one sample!"

The warning is well made. Considerable variation in grain angles around circumferences and along logs has been widely reported, even amongst those species for which regular patterns of spirality have been claimed [e.g., Harris (1965) and Pawsey (1965) for *P. radiata*; Misra (1939) for *P. roxburghii;* Paul (1956) for *P. ponderosa,* and Ozawa (1973) for *Larix kaempferi*]. Examination of weathered stumps and logs often reveals the complexity of grain patterns, especially in parts of the stem affected by roots (Fig. 2.4) and major branches.

2.2.2 Non-Destructive Sampling

Although predictions for entire stems based on a limited sample may be difficult, this need not rule out sampling systems entirely as long as their applications and limitations are fully understood. Samples from the bottom log are most accessible, and this is usually by far the most valuable log in any tree. The precision required of an estimate will depend on the effects of cross grain on wood properties required for end uses. These are discussed in detail in Chapter 3. However, it can be noted at this point that loss of strength is most marked in tensile testing, yet tensile strength is reduced by more than 10% only when grain angles exceed 5°. In this respect, therefore, it would be reasonable for some uses to aim at assessing maximum spiral angle to within ±2.5°. Extrapolations for logs other than the one from which samples are taken will be subject to greater error, and the cost of taking samples from more than one level may have to be considered. The essential first step is to test different sampling systems and to select the one that proves most cost effective, providing estimates of spirality within acceptable limits of error for relevant stem sections.

There are three main methods of obtaining samples without felling trees. The first uses observations outside the stem such as measurements made on branches. The second uses observations of grain angles in the outermost wood immediately under the bark, and the third uses wood samples extracted from the stem in ways that will (it is hoped) not endanger the tree.

Observations of features outside the stem have tended to be used as general indicators rather than for precise estimates of stem spirality. Spiral grain can be seen in roots (Krempl 1965a, c, 1970, Krahl-Urban 1967) and is said to reflect spirality in the stem. Measurement of spirality in branches of *Picea abies* (Krempl 1970) and *Larix kaempferi* (Ohkano 1984) have been directly related to stem spirality. Krempl stated that spiral grain in branches was often used by shingle makers in former times in Austria to select straight-grained trees for their purpose. In his own studies Krempl showed that spiral grain in branches tends to be more irregular and to change from LH to RH spiral earlier than that in the stem.

Ohkano (1984) made detailed measurements of 11 trees of *Larix kaempferi, L. gmellinii,* and their hybrids and found that mean values for spirality and their standard deviations were both twice as great in stem wood as in branches. However, skewness of some distributions made him question the usefulness of branch wood for predicting stem values in certain cases.

Fig. 2.5. *Eucalyptus* sp. Bark patterns indicate moderate spiral angle in the stem, increasing in the larger branches and becoming extreme in the smaller branches, e.g., *centre right.* (Photo H. G. Hiemcke-Hemming)

If there may occasionally be a useful relationship between spiral grain in branches and the trunk of some softwoods, no similar feature has been described for hardwoods. Northcott (1957) particularly observed that branches of *Alnus rubra* spiralled equally to the left or to the right regardless of the spiral in the main stem. Figure 2.5 depicts a eucalypt with moderate spirality in the stem which increases in the large branches and becomes extreme in the smaller branches.

Methods of examining grain angles under the bark have already been described. Dye injection (Sect. 2.1.3.3), radioactive tracers (Sect. 2.1.3.4), and the scribe test (Sect. 2.1.3.5) can all be used, but only radioactive tracers do not require windows to be cut in the bark. The trace produced by transpiration flow tends to dilute and broaden with increasing distance from the point of injection of dye or tracers, whereas the scribe method permits accurate measurement at all sampling points.

Removal of stem samples allows measurements to be made across an entire diameter and thus improves predictions for the logs from which they are taken. Early methods of sample extraction using special saws (as described by Dadswell and Nicholls 1959, for example) provided convenient rectangular samples useful for many types of observations (Schmidt and Smith 1961), but naturally left behind major wounds even after careful post-operative restoration. Large motor-driven hollow augers, or increment borers from 5 to 11 mm diameter, have also been used to obtain samples, and many ingenious methods have been devised to orientate their direction and to ensure clear entry into the stem (e.g., Wheeler 1964), but a full description of all these systems lies outside the scope of this book. However, as with all measurements, it is essential that processing of samples meets the requirements for correct orientation within the stem, accurate measurement, and relevant summation, which will be covered in the next section.

2.3 Methods of Measurement and Analysis

It is apparent that spiral grain does not exhibit the same regularity as other wood properties such as density or fibre length. The "normal" pattern within trees is often difficult to define, as are the sources of deviations from the pattern. All the more reason, therefore, to ensure that measurements are precise, adequate in number, and distributed within the stem in such a way as to provide the required information. It is not possible to formulate a rigid set of rules or a standard procedure for measuring and analysing spiral grain because of the different characteristics of tree species and the different ends to which measurements will be applied. However, whatever the timber, and whatever the application, it is necessary to be aware of the problems involved and the methods available to overcome them.

2.3.1 Units of Measurement

Although it is customary to measure angles in degrees, and this is the unit most frequently used for research on spiral grain in trees, there are circumstances in which other forms of measurement may prove to be more convenient. For instance, slope of grain on the surface of sawn timber may most readily be measured as a ratio, expressed as 1 in 10 or 1 in 15, for example. For conversion from degrees to the ratio form, the cotangent of the angle is used, e.g., the cotangent of 5° is 11.43, so that a slope of grain of 5° is 1 in 11.43 when expressed as a ratio. The ratio may also be expressed as a percentage, and all three systems are compared in Table 2.2.

Other systems that have been used from time to time involve the length over which a full turn of spiral is completed. Rault and Marsh (1952) obtained a value that they called "mean spiral length" by dividing the total height of a tree by the number of spiral turns along it. Krogh (1952) employed the reverse concept by

Table 2.2. Equivalent values of slope of grain expressed in degrees, ratio form, and percentage. (From Noskowiak 1959)

Degrees and minutes	Ratio form (1 : x)	Percentage
1°00′	57.29	1.75
1°55′	30.00	3.33
2°00′	28.64	3.49
2°17′	25.00	4.00
2°43′	20.00	5.00
3°00′	19.08	5.24
3°11′	18.00	5.56
3°48′	15.00	6.67
4°00′	14.30	6.99
4°45′	12.00	8.33
5°00′	11.43	8.75
5°43′	10.00	10.00
6°00′	9.51	10.52
6°20′	9.00	11.11
7°00′	8.14	12.29
7°07′	8.00	12.50
8°00′	7.12	14.04
8°08′	7.00	14.29
9°00′	6.31	15.85
9°28′	6.00	16.67
10°00′	5.67	17.64
11°00′	5.14	19.46
11°19′	5.00	20.00
12°00′	4.70	21.28
15°00′	3.73	26.81

measuring the number of spiral turns per unit of length (in this instance the length of a telephone pole or 6 m). In either approach the actual grain angle will depend on diameter of the stem in question, and therefore this type of measurement does not lend itself to direct conversion to degrees.

2.3.2 Precision of Measurement

The word precision in this section refers to the statistical precision or reproducibility of mean angles rather than to the smallest unit that can be read using a microscope, protractor, or goniometer.

2.3.2.1 Axis of Reference

The first issue to be considered when measuring an angle is the axis of reference. It is generally agreed that grain angle refers to the angle between the longitudinal wood elements and the axis of the stem: but should this refer to the initial stem axis, as indicated by the line of the pith, or to the final stem axis – the axis of the log which is the unit for commerical wood production? This issue has been argued with some heat by protagonists on either side.

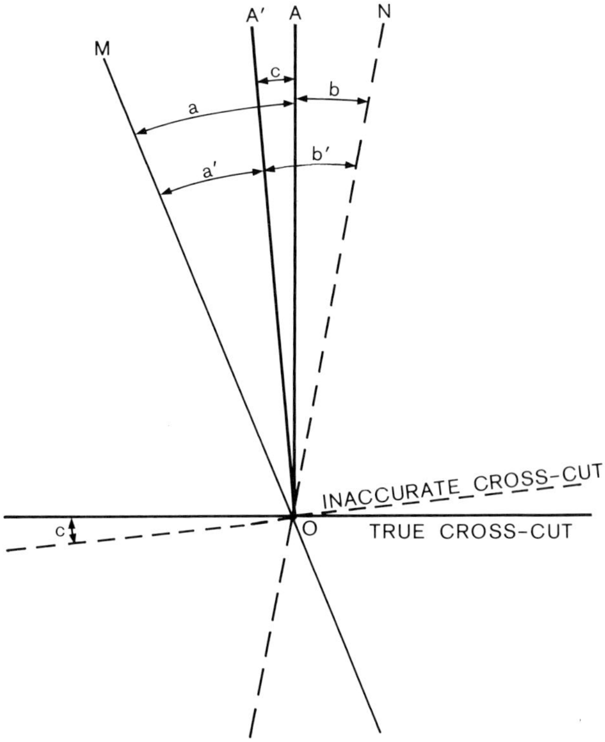

Fig. 2.6. Spiral grain measured as the average of readings on opposite radii is independent of angle of cross-cut. Grain inclination is represented by *OM* (near radius) and *ON* (far radius) at angles a and b respectively to the true axis *OA*. The mean spiral angle is $(a+b)/2$. If there is an error in cross-cutting (angle c), so that OA' is the apparent tree axis, the apparent mean spiral angle is $(a'+b')/2$. But $a+b=a'+b'$ so that the mean value is independent of the angle of cross-cut. (After Brazier 1965)

In favour of using the pith as the axis of reference (see, e.g., Northcott 1957) it can be said that this is the axis about which grain angles (and other relevant features such as internodal length) are initiated, and that it can be measured with great accuracy.

The counterargument in favour of using the log axis (see, e.g., Brazier 1967) is that this provides the angles of practical significance. Though it is conceded that the log axis may be less easily determined, any inaccuracy in defining it can be nullified by measuring angles along two radii of the same diameter and taking the averages of the pairs of readings for each annual growth layer, as explained by Brazier (1965) and illustrated in Fig. 2.6.

Brazier pursued this matter further to consider differences in spiral angle between two diameters at right angles. He found that some differences do occur, but concluded that, provided care is taken to avoid abnormal grain inclination due to factors other than spiral grain, the two sets of values give closely comparable results ($r = 0.90$) in young coniferous stems. He therefore suggested that it is more important to select a diameter free of growth characteristics that are likely to influence the general pattern of inclined grain than to work on diameters allocated by fixed cardinal points.

Elliott (1967) suggested that there is a need to distinguish between the objectives of particular studies in the choice of a datum line. Studies seeking to define the causal relationships governing changes in the cambium should logically be based on the pith axis, whereas those concerned with the practical significance of spiral grain to timber might be based on the tree axis. However, one could argue that the "axis of reference" for wood formation changes during the life of a tree. A sinuous leading shoot is often encased by cylindrical growth. The pith retains its irregular direction within this cylinder, but grain angles eventually resolve themselves into a regular pattern around the mature stem. When grain angles are plotted relative to the log axis for opposing radii across a diameter, it can often be seen that apparent differences in grain angle initiated by wandering pith move towards comparable angles in outer wood (see, e.g., Harris 1965).

There is one possible disadvantage that should be mentioned if mean values are derived for corresponding growth layers in opposite radii. Where values differ widely for reasons other than inaccuracy in defining the true axis, i.e., when there are real differences in spiral angles across opposing radii, the mean value will tend to disguise the maximum grain angles that are present, and there are applications for which maximum grain angle has more relevance than the generalised pattern of spirality.

2.3.2.2 Techniques of Measurement

Irrespective of the equipment used and the sampling patterns that are employed, precision of measurement will depend on sample preparation and operator efficiency (Balodis and Sheard 1959). In this context Burley et al. (1967) described very thorough tests to assess the precision of measured values. Two independent series of *Pinus patula* samples were prepared with the sides planed to 12 and 30 mm squares respectively on the tangential face. Five samples in each series were chosen to have grain angles representative of the range 1–8°, and each sample was assessed by five operators. To ensure adequate randomisation and independence of assessments, the experimental design used was a 5×5 latin square in each series. Samples representing different grain angles were confounded with columns, and five times of day were confounded with rows. Operators represented treatments in the classical design. Finally each latin square was replicated ten times, representing further times of day. Samples were measured to the nearest 0.25° using a modified protractor. No scribe or sliver tests were applied because the grain was sufficiently distinct in the air-dried samples.

With ten repeated observations of each sample, an operator precision of up to ±0.25° was obtained. However, operators differed significantly in their estimates of mean grain angle, although they were reasonably consistent. There was statistically significant interaction between operators and grain angles measured. It may be advisable, therefore, for one operator to be responsible for all measurements throughout an entire research project. When the most consistent operator evaluated a second, more extensive, series of samples, an overall precision of ±1° was achieved for one measurement of each sample.

For a given number of replicate measurements per operator, better precision was obtained using 30-mm rather than 12-mm wood samples. Nevertheless, the best operator obtained coefficients of variation approximating 13% even with the 12-mm samples. Thus the idea of using 12-mm increment cores for visual assessment is a distinct possibility, though microscopic evaluation would be the more usual method with this specimen size.

2.3.3 Measurements on Trees and Roundwood

Various pieces of equipment have been designed to define the axis of standing trees and logs. Fielding (1967) applied to the stem an aluminium sheet bent into a right angle. With this aligned to one side of the stem, measurements were made using a perspex scale set into one face. The scale could be adjusted laterally to measure grain angles at windows cut into the bark. When the apparatus was applied to the LH side of the stem, taper in an internode tended to cause overestimation of LH spiral angles amounting on average to 20 min of arc.

Figure 2.7 illustrates the device used by Wellner (1960) for finding the axis of a log. Two sets of legs straddle the log, and points bisecting the angle between each set of legs can be adjusted in length to indicate the true log axis. Once the axis is defined, the scribe test can be used to indicate grain angle, which can be measured by the ratio method or by protractor.

These types of apparatus would be too cumbersome for some situations, but a system described by Paterson (1967) lends itself to measurements made at various heights above ground while climbing standing trees. Paterson used an engineer's protractor to "shoot" the top of the tree, or the log axis if the stem were sinuous. Windows were cut into the bark using a 20-mm punch, and it was claimed that experienced operators could measure grain angles with considerable precision using this equipment, and that "very acceptable" agreement was achieved between operators.

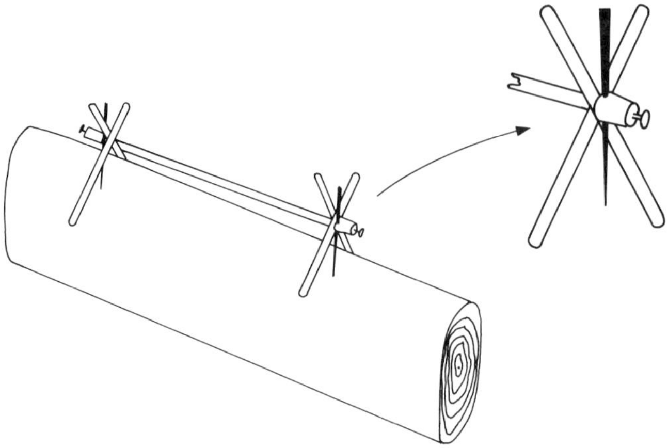

Fig. 2.7. Device for defining a log axis. (After Wellner 1960)

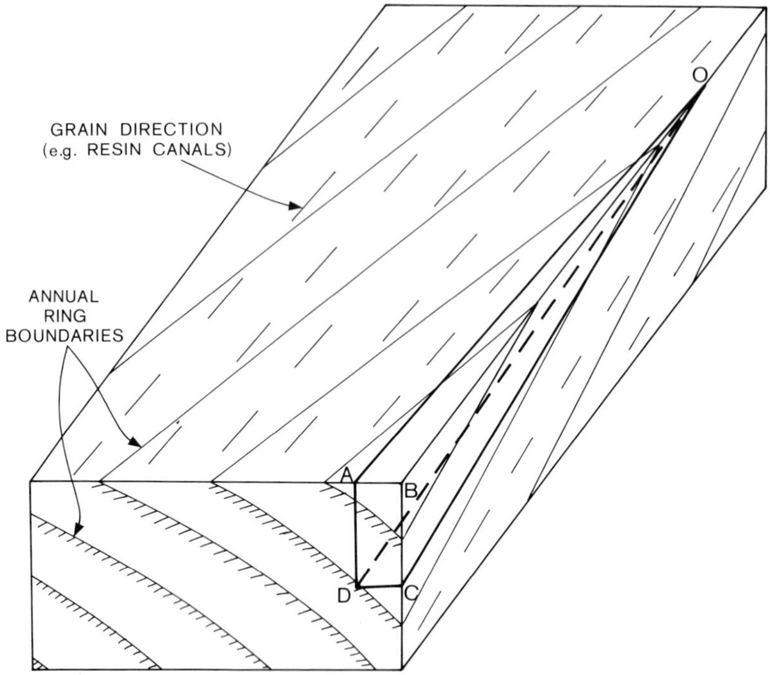

Fig. 2.8. Measuring true slope of grain in sawn timber. Combined slope of grain *OD*:*OB* can be calculated from measurements of slopes of grain on two faces at right angles to each other, *OA*:*OB* and *OC*:*OB*. (After Anderson et al. 1945, US For Serv FPL, Madison WI)

2.3.4 Measurements on Sawn Timber

Sawn timber seldom has faces that are exactly tangential and radial. Grain angles on any surface may therefore reflect many sources of grain deviations, and their combined effects are often obscure. To calculate the true slope of grain within a piece, the grain angle on two faces at right angles should be measured as shown in Fig. 2.8. Measurements need not, of course, be made from the same point on a corner. This position has been used in the figure to clarify the geometry of measurement. However, since grain directions may vary considerably along a piece, it is desirable for the two measurements to be made at places close enough to avoid these variations.

If slope of grain on the wide face is expressed as the ratio AB : BO, and that on the narrow face as BC : BO, the combined slope of grain DO is represented by BD : BO and can be determined by reference to Table 2.3.

Several instruments suitable for measuring grain angles on sawn timber have been described by Anderson et al. (1945). However, for most purposes a simple pair of dividers will be quite adequate to measure, firstly, a suitable distance along a baseline drawn parallel to one edge, and then the deviation of the grain from it.

Table 2.3. Combined slope of grain[a]. (Anderson et al. 1945, US For Serv FPL, Madison WI)

To determine the combined slope find the column marked by the steeper of the two slopes and in this column locate the figure in line with the less steep of the two slopes as given in the left-hand column. For example, if the slope on the face is 1 in 15 and that on the edge is 1 in 20, the combined slope is 1 in 12.0

Slope	1 – 10	1 – 11	1 – 12	1 – 13	1 – 14	1 – 15	1 – 16	1 – 17	1 – 18	1 – 19	1 – 20	1 – 21	1 – 22	1 – 23	1 – 24	1 – 25	1 – 26	1 – 27	1 – 28	1 – 29
1 – 10	7.1																			
1 – 11	7.4	7.8																		
1 – 12	7.7	8.1	8.5																	
1 – 13	7.9	8.4	8.8	9.2																
1 – 14	8.1	8.7	9.1	9.5	9.9															
1 – 15	8.3	8.9	9.4	9.8	10.2	10.6														
1 – 16	8.5	9.1	9.6	10.1	10.5	10.9	11.3													
1 – 17	8.6	9.2	9.8	10.3	10.8	11.2	11.7	12.0												
1 – 18	8.7	9.4	10.0	10.5	11.1	11.5	12.0	12.4	12.7											
1 – 19	8.8	9.5	10.1	10.7	11.3	11.8	12.2	12.7	13.1	13.4										
1 – 20	8.9	9.6	10.3	10.9	11.5	12.0	12.5	13.0	13.4	13.8	14.1									
1 – 21	9.0	9.7	10.4	11.1	11.6	12.2	12.7	13.2	13.7	14.1	14.5	14.9								
1 – 22	9.1	9.8	10.5	11.2	11.8	12.4	12.9	13.5	13.9	14.4	14.8	15.2	15.6							
1 – 23	9.2	9.9	10.6	11.3	12.0	12.6	13.1	13.7	14.2	14.6	15.1	15.5	15.9	16.3						
1 – 24	9.2	10.0	10.7	11.4	12.1	12.7	13.3	13.9	14.4	14.9	15.4	15.8	16.2	16.6	17.0					
1 – 25	9.3	10.1	10.8	11.5	12.2	12.9	13.5	14.1	14.6	15.1	15.6	16.1	16.5	16.9	17.3	17.7				
1 – 26	9.3	10.1	10.9	11.6	12.3	13.0	13.6	14.2	14.8	15.3	15.9	16.3	16.8	17.2	17.6	18.0	18.4			
1 – 27	9.4	10.2	11.0	11.7	12.4	13.1	13.8	14.4	15.0	15.5	16.1	16.6	17.1	17.5	17.9	18.3	18.7	19.1		
1 – 28	9.4	10.2	11.0	11.8	12.5	13.2	13.9	14.5	15.1	15.7	16.3	16.8	17.3	17.8	18.2	18.6	19.1	19.4	19.8	
1 – 29	9.5	10.3	11.1	11.9	12.6	13.3	14.0	14.7	15.3	15.9	16.5	17.0	17.5	18.0	18.5	18.9	19.4	19.8	20.1	20.5

[a] Based on formula: Combined slope of grain = $(\text{slope on face})^2 + (\text{slope on edge})^2$ where slope of grain is expressed as fraction.

For example: $\left(\frac{1}{15}\right)^2 + \left(\frac{1}{20}\right)^2 = \frac{1}{12}$ or 1 in 12.

2.3.5 Measurements on Large Wood Samples

Specimens larger than those normally extracted using an auger or increment borer (e.g., measuring more than 25 mm along the grain) may be cut from felled trees and logs as cross-sections, or sawn from standing trees using special techniques. Diametric or radial strips prepared from these can be examined either by sectioning the sample at intervals and measuring grain angle on tangential surfaces, or by splitting the whole section radially.

Splitting is a method that has been favoured by many workers, including several who have used quantitative analytical techniques (Northcott 1957, Kennedy and Elliott 1957, Elliott 1958, Nakagawa 1972, Mikami and Nagasaka 1975). Grain angles across a radius can be measured directly using a modified protractor (Northcott 1957), or by measuring deviations from a projected line through the pith (Nakagawa 1972). However, Miyaki et al. (1983) found that where the axis of the pith is not in line with the stem axis, deviations occur about the projection line so that the split radii create a V shape. Therefore they recommended that average values for two opposite radii be used to counteract this effect. It should also be noted that some species, e.g., *Araucaria cunninghamii*, (Balodis and Sheard 1959) apparently do not split accurately with the grain, so that no assumptions should be made, especially when examining an unfamiliar timber.

If the pith is not being used as the axis of reference, measurements across a radius can only be as accurate as the defined axis. There are several ways of tackling this problem. A plane surface along the radial longitudinal or transverse face of the sample, accurately machined to be parallel or at right angles to the log axis may be used to provide the axis of reference. For example, the transverse faces of discs cut at right angles to the log axis have been used by Balodis and Sheard (1959) and Kromhout (1966). Alternatively, the grain angle at the outer surface of the sample can first be measured relative to the log axis by any of the methods described in Section 2.3.3, and the angle between this and an arbitrary plane along the sample radius can then be used to calibrate all interior measurements made relative to the plane surface (Paterson 1967). Another method is to relate all interior angles directly to the longitudinal axis as marked on the outer surfaces before removing the sample (Gerischer and Kromhout 1964, Hughes and Andrew 1974).

2.3.6 Measurements on Increment Cores

Many wood properties can be reliably assessed from increment cores (e.g., Keller 1973), but several other workers have questioned whether or not spiral grain can be accurately determined from 5-mm increment cores. Phillips (1963) described experiments in which the core extractor was inserted with its edges at right angles to the stem axis, in the hopes that the score marks left on the core could be used to provide a reference axis. Unfortunately this did not prove feasible because the extractor, made of spring steel, often twisted during insertion. Noskowiak (1968) found that the cores themselves often became permanently twisted, and he therefore used a larger (12-mm diameter) borer which he found did not produce distortion in the sample.

Quite obviously, the strain exerted by frictional and other forces within the borer must not exceed the elastic limit in torsion for the wooden core being extracted. Since friction increases directly with core diameter, and torsional strength with the square of the diameter, a large core is less likely to be deformed than a small core, other things being equal.

It may be possible to examine species with high torsional strength using a 5-mm borer, and Keller (1973) recorded no problems in using a borer of this diameter to examine *Pinus pinaster*. However, Harris (1984) concurred with Noskowiak (1968) in preferring to use a borer of 12-mm diameter for examining spiral grain.

Noskowiak (1968) used a rather complex system to measure grain angles along a core. The grain angle at the outer wood surface relative to the stem axis was determined before the core was extracted. Then, after extraction, annual ring widths were accurately measured before the whole core was sealed into a tube of

Fig. 2.9. Apparatus to measure grain angle relative to vertical in a standing tree. (Photo H. G. Hiemcke-Hemming, from Harris 1984)

rectangular cross-section with beeswax. The tube was then mounted in a microscope, and grain angles were measured at 40×magnification, under incident lighting using a goniometer eyepiece. The sample embedded in beeswax was progressively extruded by amounts equivalent to the measured annual ring widths using a compressed air device, and the extruded portion was severed to provide the next tangential surface for measurement. Grain angles measured on interior growth layers were compared with those at the outer wood surface for conversion to values relative to the stem axis.

Harris (1984) used a somewhat simplified method to obtain average values for measurements of each growth layer on opposite radii of a 12-mm core. Grain angles on the outer faces of a diametral core were measured, before extraction, using the apparatus in Fig. 2.9. The two points on the apparatus are inserted into a scribe mark on the wood surface, and the straight edge of the protractor is rotated to the horizontal using the spirit level. Measurements of grain angle are therefore made relative to true vertical. The average of two readings from opposite radii will then give the mean slope of grain relative to the stem axis, irrespective of any stem lean, as explained in Section 2.3.2.1 and illustrated in Fig. 2.5 (the hypothetical "cross-cut" in this instance being always horizontal, and not necessarily at right angles to the stem axis).

After extraction, a plane surface is cut arbitrarily down one side of the core, and grain angles at the outer surface of each annual growth layer are measured microscopically relative to this surface using incident lighting. Two angles have thus been measured on the outermost faces of the core: before extraction, grain angles were measured relative to true vertical, and after extraction, relative to the arbitrary plane surface. The differences between these two sets of readings can now be applied to measurements relative to the plane surface for the other growth layers to convert them to relativity with true vertical.

2.3.7 Measurements on Veneers

Woodfin (1969) ingeniously used the progressive disclosure of internal growth layers as a log is peeled for veneer to examine spiral grain in *Pseudotsuga menziesii.* Samples were taken at radial intervals of 38 mm (1.5 inches) to provide approximately 9000 pieces of veneer representing 1845 peeler blocks and 167 trees. Each green veneer sample was split longitudinally and grain angle and direction were measured to the nearest degree. Spiral grain patterns were coded using the system devised by Lowery (1966).

Woodfin was also one of the first to use a computer programme to provide weighting relative to the volume represented by each veneer sample, as will be described in the next section.

2.3.8 Analysis of Measurements

The numerous way in which grain angles can vary throughout a stem make any simple form of summation very difficult. In a purely descriptive sense, several

CLASSIFICATION OF GRAIN PATTERNS

A (AFTER LOWERY 1966)

SPIRAL GRAIN PATTERN CODE	DESCRIPTION
1	LEFT INCREASE AWAY FROM ZERO
1a	LEFT INCREASE, THEN DECREASE TOWARD ZERO
2	LEFT DECREASE TOWARD ZERO
2a	LEFT DECREASE THEN INCREASE AWAY FROM ZERO
3	CHANGE FROM LEFT TO RIGHT SPIRALITY
4	LEFT SPIRAL WITH LITTLE CHANGE
5	RIGHT INCREASE AWAY FROM ZERO
5a	RIGHT INCREASE, THEN DECREASE TOWARD ZERO
6	RIGHT DECREASE TOWARD ZERO
6a	RIGHT DECREASE, THEN INCREASE AWAY FROM ZERO
7	CHANGE FROM RIGHT TO LEFT SPIRAL
8	RIGHT SPIRAL WITH LITTLE CHANGE
9	ESSENTIALLY STRAIGHT GRAIN

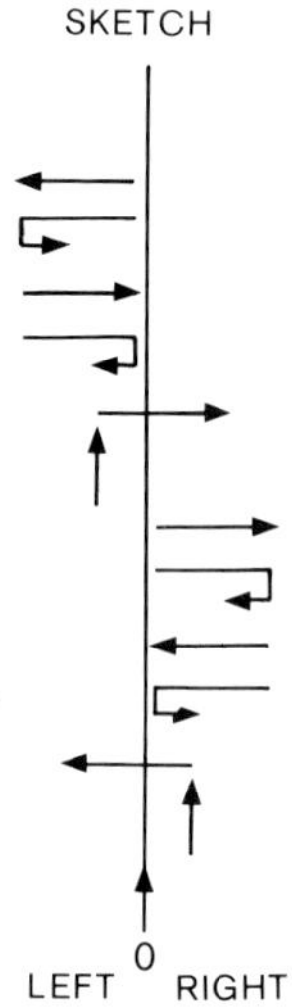

B (AFTER OHKURA 1958)

1 STANDARD TYPES OF SPLITTING CURVE

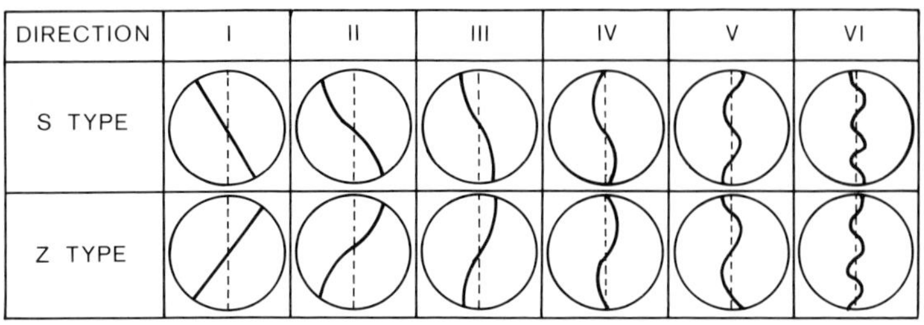

2 MISCELLANEOUS TYPES

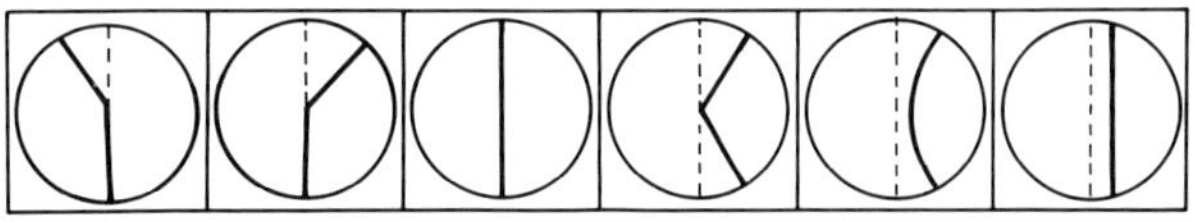

Fig. 2.10. Two methods of classifying grain patterns

workers have devised systems to indicate the general grain patterns that may be encountered. Two are illustrated in Fig. 2.10.

Lowery (1966), who suggested the first of these, was concerned with spiral grain in conifers, and his code lends itself quite well to describing the main patterns in this category. Ohkura (1958), on the other hand, required a system that would describe the splitting curves of a great variety of hardwoods and softwoods. His patterns therefore show a dotted line to indicate the diameter where splitting is initiated on one face of a cross-section, and a solid line to indicate the split surface that results.

For many purposes, however, it is more desirable to quantify spiral grain so that direct comparisons can be made between trees, to allocate logs to most suitable uses, to select trees for breeding, or to distinguish between populations. Lowery (1966), for example, examined the frequency distribution of his spirality patterns in different populations of *Pinus contorta, Picea engelmannii, Pseudotsuga menziesii,* and *Larix occidentalis* and came to the conclusion that patterns were unaffected by geographical location.

Two methods have been proposed to summarise grain angles as a single statistic for each tree. Kennedy and Elliott (1957) devised what they termed the "cumulative absolute spiral angle". In this expression the direction of spirality is ignored on the grounds that only the absolute degree of spirality affects timber properties. Furthermore, the absolute spiral angle at any point was plotted as a cumulative figure, representing the total number of degrees through which the spiral had progressed from the pith. As an example, they cited a tree which had 4° RH spiral at 10 years and changed steadily to 5° LH spiral at 50 years. The absolute spiral angle for this tree would be 9° at 50 years.

Brazier (1965) criticised cumulative absolute spiral angle because it gives undue weighting to values near the pith which affect a relatively small volume of wood. One might add that a tree with relatively straight grain, fluctuating between 2° LH and 2° RH at, e.g., 5-year intervals, could pick-up an absolute spiral angle of 40° at 50 years.

To provide some indication of the volume of wood affected by spirality, Brazier (1965) proposed a value based on successive observations across a stem weighted according to the proportional volume represented by each reading. This value he termed the "spiral grain index". Other workers have extended the use of weighted values to include weighting for log volumes in each stem as well as for volumes of increments in each log (Balodis 1960, Whyte et al. 1980). Woodfin (1969) used a similar approach to produce what he termed "weighted average spirality" for grain angles measured on veneers.

The use of a single statistic has been criticised by Nicholls (1963) who believed that between-tree comparisons are more effective using curves of variation from pith to bark. Commenting on this suggestion Burley et al. (1967) stated that where no consistent pattern of spirality can be established between radii, heights, or trees, a mean value may justifiably be used to represent a tree. However, they went on to say that if a systematic pattern of variation can be identified, the mathematical parameters of this pattern should be determined and compared between trees. There are various possible approaches to the representation (and hence parameter determination) of such patterns.

If actual data follow a well-defined pattern, the best curve may be fitted. An example is the analysis by W. G. Warren which appears as the appendix to Harris (1965). Grain angles for each annual ring, in each annual height increment, were compared for three 7-year-old trees in each of five clones of *Pinus radiata.* Cubic polynomials were fitted to the radial patterns of grain angles, and indicated that the patterns (though not necessarily the absolute values) were more uniform within than between clones.

When the relationship between grain angle and ring number is less distinct, a better impression may be gained by using moving point or cumulative averages (e.g., Gerischer and Kromhout 1964, Burley et al. 1967).

2.4 Measurement of Interlocked and Wavy Grain

Interest in measuring interlocked grain derives largely from those species in which is not regarded so much as a decorative feature but rather as a defect giving rise to problems in machining and to warping in sawn timber and veneers. Despite its apparent complexity, the few studies that have attempted to describe and quantify interlocked grain have been quite successful in a pragmatic sense.

Martley (1920) first described what he called the "amplitude" and "period length" of splitting curves for blocks of standard length (Fig. 2.11), and the same general procedure was used by Limaye (1954) for a study of Indian woods. He also used the term amplitude, but referred to the distance between successive maxima in the same direction as the "wavelength". Limaye then compiled a table setting out average amplitudes and wavelengths for seven intensities of interlocked grain, so that different species could be assessed in relation to these standards.

Webb (1967, 1969) also used split samples of standard length to study interlocked grain in *Liquidambar styraciflua.* He assessed the degree of interlocking by measuring the areas of deviation about the projected splitting line on graph paper with 1-mm squares. The sums of such areas were then divided by the radial width in centimetres to provide an "interlocked grain index". This method was adopted by Krawczyszyn and Romberger (1980). Webb was able to demonstrate the significance of this index in two respects. Indexes calculated for samples from breast height proved to be significantly correlated ($p<0.01$) with the indexes for entire stems ($r = 0.71$). In addition, the index for any sample was strongly correlated ($p = 0.01$) with the maximum angular deviation within the sample ($r = 0.84$).

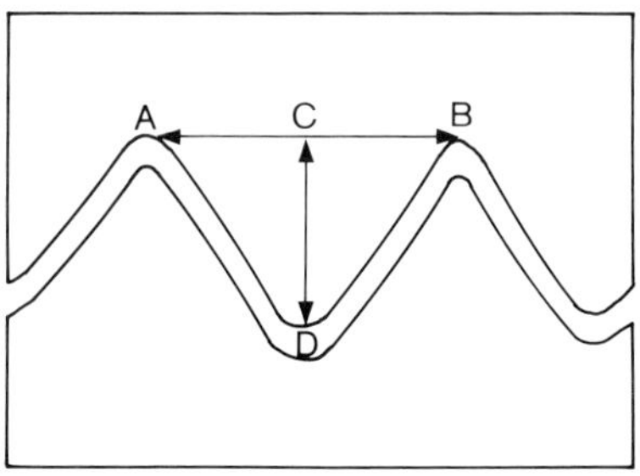

Fig. 2.11. Descriptive parameters for interlocked grain using the splitting curve for a wooden block of standard length. (After Martley 1920, Limaye 1954). *AB* Period length (Martley) or wavelength (Limaye); *CD* amplitude (both authors)

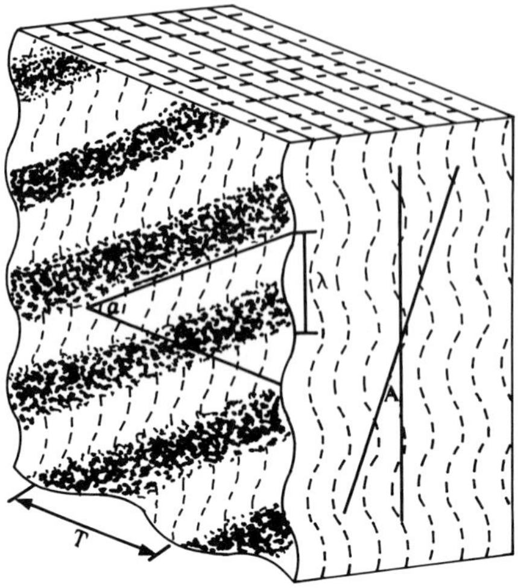

Fig. 2.12. Three-dimensional model of wood with wavy grain. *A* Amplitude; *λ* wavelength; *α* angle between bands of different light reflectance and wood rays; *T* period. (After Pyszynski 1977a)

Noskowiak (1959) expressed some concern about the use of blocks with standard dimensions to assess grain deviations in that no mention is made of variability in the depth to which the splitting blade cuts through the wood before true splitting is initiated. From the results described above it would appear that a standard technique can be developed, although this feature is not discussed by any of the writers.

The relative rarity or intermittent occurrence of wavy grain usually lends to it an aesthetic, or at least curiosity value, so that it is less frequently regarded as being a defect. When grain follows a regular wave pattern it can be described in terms of amplitude, wavelength, and period, as illustrated in Fig. 2.12. Of more complex figures it can only be said that they would appear to defy measurement, and often beggar description.

2.5 Conclusions

1. Grain angles tend to show considerable variability within stems, and deviations about any generalised pattern are greater than those for other wood properties such as cell dimensions or wood density. In some species spiral development appears to be almost completely random.
2. There are many methods of detecting and measuring spiral grain. Selection of the most appropriate process will depend on the type of study to be undertaken.
3. No set systems of measurement or analysis can be recommended because both will depend on
   (a) the pattern, intensity, and variability of spiral development;
   (b) the effect of grain angle on physical and mechanical properties (as will be discussed in Chap. 3); and
   (c) the interaction between both of these aspects for the desired timber uses.
4. Development of methods most appropriate to any study of spiral grain requires complete understanding of all these features.

Chapter 3

Effects of Grain Angle on Wood Properties and Uses

3.1 Consequences of Anisotropy

Northcott (1965) made the point that if wood were isotropic, spiral grain would have no influence on its properties. It is because wood has different properties in the longitudinal, radial, and tangential directions relative to the orientation of its cellular components that any change in cellular orientation affects its usefulness. Except for the most extreme forms of spiral grain – with grain angles in excess of 30°, for example – the effects of spirality appear to relate solely to changes in the axes of anisotropy. Thus the magnitude of the changes which result from a given level of spirality depends on the degree of anisotropy normal for the species. In this connection it may be noted that the anisotropy of coniferous wood is much more pronounced than that of broad-leaved woods (Kollmann and Côté 1968) and therefore properties of softwoods should change more than those of hardwoods in response to spiral grain. However, the effects of anisotropy may differ quite markedly even between coniferous woods. For example, Banks (1969) noted that *Araucaria columnaris* twists on drying less than *Pinus patula* with the same spiral angle.

The anatomical structure of wood with low to medium grain angles is identical with that of straight-grained wood. Only at extreme grain angles is wood structure significantly affected, in ways that probably reflect the basic need for water to move along the stem. Where grain angles approach the horizontal in *Pinus sylvestris*, for example, tracheids have been observed to have more bordered pits, distributed along their entire lengths rather than being concentrated towards the ends (Hartig 1895, Liese and Ammer 1962), thus facilitating movement across them and along the stem. Even so, other changes are slight. Liese and Ammer found that the normal radial increases in tracheid length and ray height were retarded when spirality exceeded 30°, and that earlywood tracheid walls were slightly thicker. However, no changes were observed in tracheid walls of latewood, in diameters of lumina in either earlywood or latewood, or in frequency of rays per unit area.

3.2 Effects on Shrinkage

Harris (1965) confirmed that wood with spiral grain has no abnormal shrinkage properties per se, but gives rise to abnormal shrinkage patterns in sawn timber because a component of the tangential shrinkage (at right angles to the longitudinal wood elements) is translated into the longitudinal direction of the timber (Boas 1933, Paul 1957).

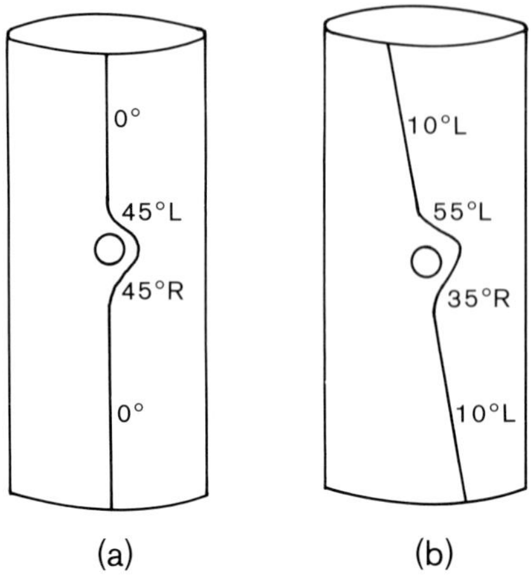

Fig. 3.1. Spiral grain may supplement maximum grain deviations from other causes. Diagram of grain deviations around a branch in a straight-grained tree **(a)** and a tree with 10° LH spiral grain **(b)**

In practice it is difficult, and often pointless, to attempt to isolate the sources of cross grain in timber, so that it need only be remarked that the effects of grain angle on shrinkage will apply equally whatever their origins may be. However, spiral grain commonly affects the entire length of a piece of sawn timber, whereas grain deviations due to knots, nodal swellings, and small stem kinks tend to be more localised. Moreover, spiral grain tends to augment the effects of other deviations when these lie in the same direction, so that maximum grain angles are increased. For example, grain deviations around a branch develop relative to the general grain direction in that region, so that spiral grain adds to the maximum deviation in the same direction as the spiral angle, as shown in Fig. 3.1.

3.2.1 Warping in Sawn Timber

Any distortion about the original surface of a piece of wood as a result of a change in moisture content is referred to as warping. The type of warping arising from spiral grain is called twist, which develops as the turning of a plane surface along its length. It is possible to predict, to a certain degree, how a piece of timber will shrink and warp in drying by examining the orientation of the annual rings and the grain direction (Kollman and Côté 1968).

The way in which spiral grain induces twist is illustrated in Fig. 3.2. A flat-sawn board cut from a spirally-grained log will have grain running in opposite directions on opposing edges. The direction of grain on the face of the board will change about a plane, through the length of the board, that is both radial to the growth rings and at right angles to the board surface (Brazier 1965).

The relationship between twist and grain angle can be derived theoretically for a cylindrical shell (Stevens and Johnston 1960, Ohkura et al. 1961). If the grain direction in a hollow wooden cylinder is not parallel to the axis of the cylinder, it can be shown that when one end of the cylinder is fixed, the other end will rotate with changing moisture content. Stevens and Johnston (1960) obtained

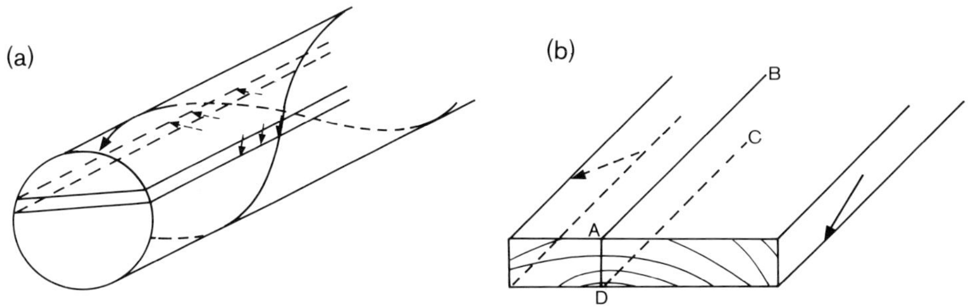

Fig. 3.2. a Opposing grain directions on the edges of a board cut from a log with spiral grain. **b** The grain inclination relative to the surface of the board changes on either side of the plane *ABCD*. (After Brazier 1965)

good agreement between experimental results with hollow wooden cylinders, and predicted values of rotation, as moisture content increased.

The equation developed by Stevens and Johnston related only to rotation in an expanding hollow cylinder. By introducing signs, Balodis (1971, 1972) changed the equation to apply to both expansion and contraction:

$$t = \pm(\theta/r)(2\,Le)/(1 \pm e) \ ,$$

where t = angle of rotation at a diameter of the free end of the shell in degrees;
θ = grain angle in degrees;
e = tangential shrinkage (or expansion) per unit width;
L = length of shell in cm;
r = radius of shell in cm.

The positive signs in this equation correspond to the rotation of a contracting shell such that rotation is in the same direction as the grain angle during shrinkage. Balodis also pointed out that the equation is strictly valid only for small grain angles because it has been assumed that $\sin\theta = \tan\theta$ in its development.

Balodis (1972) then considered the case of a plane surface parallel to the axis of the cylinder. The chord formed by the intersection of this plane and the circular cross-section of the cylinder at the free end will also undergo rotation relative to the corresponding chord at the fixed end. The analogy, of course, is between this plane and the surface of a board, so that the angle of rotation of the chord indicates the angle of twist of the board.

Stevens and Johnston (1960) pointed out that one feature of the theoretical equation is that, if the spiral angle were constant throughout a log, annual rings would have a tendency to twist decreasing in proportion to their distance from the pith. This would induce internal torsional stresses. Assuming no more than that grain angle should change only slightly across the diameter, it would follow that the larger the log diameter, the smaller would be the tendency for it to twist during drying. It would also follow that flat-sawn material from trees of large diameter (wide boards) would twist less in drying than narrow boards cut from smaller trees.

In practice, Stevens and Johnston concede that the complexity of spiral grain patterns, and the involved formation of internal stressing and "sets", virtually precludes accurate estimation of the actual twist that will develop in a drying log or board. In this respect Balodis (1972) found that the theoretical twist exceeded observed values. Among possible causes for this discrepancy, he included external restraint during seasoning, and internal restraint. In support of internal restraint he cites the fact that small wood samples exhibit higher tangential shrinkage than larger samples. The observation of Kloot and Page (1959) that distortion is reduced when large balks are pre-seasoned prior to final cutting, also supports the contention that internal restraint can significantly reduce distortion in sawn timber.

Both theoretical and practical results indicate that the main structural feature of wood associated with twist in dried boards is the ratio of grain angle to distance from the pith. For any given grain angle, twist in boards can be reduced by increasing board width or by increasing the ratio of width to thickness (Balodis 1972). For practical purposes. twist can be reduced by drying boards under constraint at high temperatures (Van de Merve 1973), and angled bolsters can be used to induce "reverse twist" to compensate for any relaxation after drying.

However, any change in moisture content after drying will also induce twist in spirally grained boards, and this probably accounts for the considerable differences in attitude towards spiral grain in different countries. Where equilibrium moisture content (e.m.c.) can vary widely between different locations within a country, timber producers become very conscious of distortions that result from shipping between locations. In South Africa, for example, e.m.c. in a narrow coastal strip is 15%, whereas over the major part of the country it lies between 6 and 10% (de Villiers 1973). Similarly, countries that experience few problems in seasoning timber to meet local conditions may face unfamiliar difficulties when exporting to different climatic zones.

The variables that determine the degree of distortion arising from spiral grain have not always been clearly identified and have given rise to some apparently conflicting results. Paul (1955) found that narrow boards cut from the outside of *Pinus clausa* logs twisted more than wide boards cut from near the centre. This was because spiral angles were highest towards the outside of these logs. In *Pinus roxburghii*, Rault and Marsh (1952) found that twist was more severe in boards cut from logs of small diameter than in those cut from large-diameter logs. In this case the small-diameter logs had more severe spiral grain.

In most young coniferous logs, however, such as those that are produced in plantation forestry, an initial LH spiral near the pith is the feature that dominates the development of twist in drying. Consequently, sawn timber containing pith exhibits the most distortion (Kano et al. 1964, 1965, Wong 1980).

Kano et al. (1965) also observed that twist increased with the area of knots measured on the outside of spirally grained squares of *Larix kaempferi*. Though grain deviations around knots would not influence the overall spiral angle, it is possible that by increasing the maximum spiral angle, as described in Section 3.2, they intensify the effect of spiral grain in critical areas.

3.2.2 Movement in Roundwood

Twisting of tree trunks and roundwood with spiral grain is sometimes quite dramatic, and its effects seen in the wild roused Gunther (1929) to a lyricism which surely demands quotation in full:

"On the northern slopes of the San Gabriel Ranges commanding the Mojave Desert, near the Smithsonian Solar Observatory, a pine (probably *P. ponderosa*) smitten by the rigor of death before it fell, continued twisting after it had fallen. The length of the trunk was some 30 feet, and the diameter at the base two and a half, a volume of timber of considerable weight. As the twisting continued when the tree lay on the ground, the weight of the base of the tree anchored it, and made it the fixed point on which the upper half of the tree revolved. The upper part had suffered a dextral twist so thorough that the branches, still far from decomposed and several inches in thickness, were snapped off like laths as they were pressed in turn laterally against the ground. Upon the right side of the trunk lay a pile of logs and kindling ready broken; on the left little loose wood, but appearing on the left underside of the trunk were the scars of the branch bases from which the branches had been broken off on the right flank. It did not seem as if the branches even slightly retarded the powerful twist of the trunk."

The general principles governing torsion in drying as set out in Section 3.2.1 also apply to roundwood such as posts and poles, in which the annual growth layers can be regarded as a number of concentric "cylindrical shells" joined together. The interaction of growth layers with different grain angles and varying moisture contents, together with the internal and external restraints already referred to, make prediction of the degree of twist that will be encountered for any change in average moisture content quite impossible. Nevertheless, twisting of poles in service which, according to Burger (1953), can be as much as 50°, is a well-known phenomenon, and the greatly magnified movement at the ends of cross-arms stresses, and sometimes breaks transmission lines. However, Moll (1947) cast doubts on the significance of such movement in practice.

Various methods have been described for measuring this twisting or "working" (Burger 1953) of poles, but in essence most observe movement of a diameter at the free end of a pole while the other end is fixed. In erect (planted) poles this can be accomplished by suspending a plumb line from one end of a cross-arm or bar at the upper end of the pole and observing movement relative to a fixed point on the ground (Krogh 1952, Burger 1953, Wellner 1960, Lowery 1967a, Lowery and Erickson 1967).

Lowery (1967a) described another method using a large protractor to measure the angle between the cross-arms and a line wire, making the assumption that the line wire was originally at right angles to the arm. Burger (1953) also made some observations on poles lying out of the ground, by marking the original vertical diameter on either end and measuring the deviations that developed as the moisture content changed.

The general observations of pole twisting can be summarised as:

1. When poles are installed with a moisture content above the shrinkage intersection point, the initial twisting as the pole dries to approximate equilibrium

with its environment far exceeds that which subsequently arises as the result of seasonal fluctuations in moisture content (Krogh 1952, Burger 1953, Keating 1966, Wellner and Lowery 1967). Drying, prior to installation, to approximately the mean local e.m.c. is therefore the most effective precaution that can be taken.

2. In conifers that follow the common pattern of spiral grain, being initially LH and sometimes changing to straight grain or RH spiral in outer wood, poles that have LH spiral grain throughout twist more than those with straight grain or RH spiral on the outside. Poles with severe RH spiral grain may twist to the right on drying, but twisting is often reduced by the contrary trend from the internal LH spiral. (Krogh 1952: *Pinus pinaster, P. canariensis*; Burger 1953: *Picea abies, Abies pectinata*; Keating 1966: *Pinus radiata, P. nigra*; Lowery 1967a: *Pseudotsuga menziesii, Pinus contorta, Larix occidentalis.*)
3. Krogh (1952) included some hardwood poles in his tests (*Eucalyptus saligna, E. paniculata*) which had RH spiral grain increasing from the centre outwards, and which therefore twisted to the right in drying. Their movement in response to fluctuating e.m.c. was less than that of the pines (even those with consistent and equivalent LH spiral grain) which Krogh suggested may reflect the lower rate of moisture absorption and loss by the eucalypts.
4. Poles treated with creosote and oilborne preservatives also gain and lose moisture much more slowly than untreated poles or poles treated with waterborne preservatives. Once they have reached equilibrium, poles treated with oily preservatives are therefore relatively stable (Krogh 1952).

3.2.3 Warping in Plywood and Sliced Wood

Brouse (1961) stated that for all practical purposes plywood construction can be considered to be under stress at all times. If stresses are balanced the panel will remain flat, but if stresses acting on one side are not sufficiently well balanced by opposing stresses on the other, warping will result.

Twisting in plywood usually arises from differences in grain direction. Brouse (1961) used the example of a five-ply panel with a comparatively thick core and thinner cross-bands and faces. The grain of the cross-band on one side should be parallel to that on the other. The variation that may be tolerated will be affected by the thickness of the core, moisture content at time of gluing, service conditions, and so on. However, it has been demonstrated experimentally that differences as little as 5° in grain direction of opposing cross-bands can cause noticeable twisting, and pronounced twisting in commercial plywood has been associated with differences of 15°.

Grain direction on the face of rotary-cut veneer will be directly affected by spiral grain. Straight-grained species are therefore greatly to be preferred for cross-banding. An alternative is to clip veneer parallel to the grain, though this is usually far too time-consuming. In special products cross-band stock can be clipped into narrow widths, reversing alternate strips end for end, and taping or edge gluing (Brouse 1961).

Banks (1971) examined the use of South African grown *Pinus strobus* for rotary-cut veneer and *P. roxburghii* for sliced box shook. He found little effect of the spiral grain that was present in both species on properties of plywood made from *P. strobus* or on banana crates made from *P. roxburghii*. However, he emphasized that the type of utilisation tended to minimise the effect of the spiral grain. Spiral grain in *P. strobus* is greatest in the centre of the log, which was discarded in rotary peeling, while spiral grain in *P. roxburghii* is greatest towards the outside of the log, and this was lost as slabwood in preparing flitches for slicing. These results serve to emphasise that both the method of utilisation and the end use may have considerable bearing on the consequences of using spirally grained logs.

3.3 Effects on Wood Strength

Strength properties of wood are strongly anisotropic, the greatest differences being between properties measured "along the grain" and "across the grain". The ratio of tensile strength parallel to, compared with perpendicular to the grain varies between 25 : 1 for green wood and 45 : 1 for air-dried material. In compression, the difference is not so marked, being only six to ten times greater parallel to the grain than perpendicular to it (Boas 1933).

3.3.1 Strength of Sawn Timber

Figure 3.3 illustrates how strength in tension, bending, and compression are affected by grain direction. It will be seen that loss of strength amounts to 10% for grain angles somewhere between 5° and 10° (1 : 11 to 1 : 6) for all three properties (Kollmann and Côté 1968). Impact strength is particularly susceptible to sloping grain and it has been suggested that slope of grain should not exceed 1 in 25, or a little over 2°, for impact uses (Wilson 1921, Koehler 1924, Boas 1933).

The loss of strength that can be tolerated for various uses is, of course, partly a matter of opinion. Timber-grading rules must make allowance for this defect, and the issue has been faced in different ways in different countries. Moreover, rules can be changed to meet particular circumstances, so there is little point in

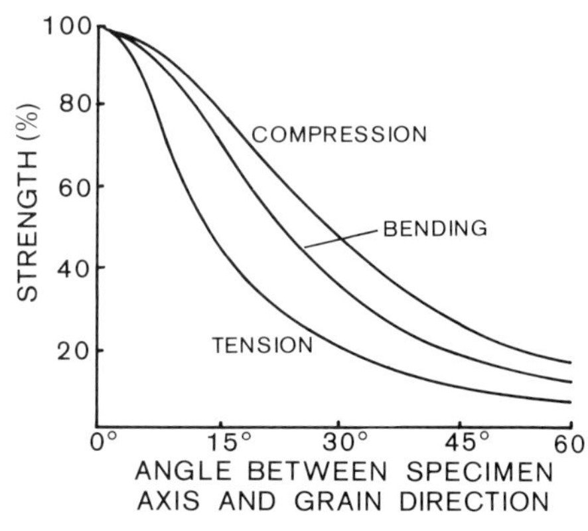

Fig. 3.3. Dependence of strength properties on the angle between grain direction and the specimen axis. (After Kollmann and Côté 1968)

Table 3.1. Maximum slope of grain permissible according to selected authors

Reference	Maximum slope of grain for wood in	
	Bending	Compression
Wilson 1921	1:20	1:10
Boas 1933	1:15	1:5
Banks 1953	1:8	
Markwardt and Wood 1958	1:20	1:15
Luo and Xu 1983	1:10	

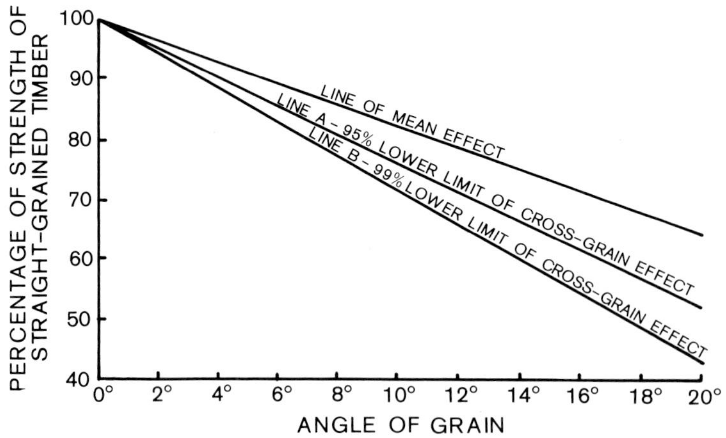

Fig. 3.4. Lower limits of cross-grain effect on modulus of rupture of green *Eucalyptus marginata*. (After Kloot and Schuster 1958)

attempting to tabulate current restrictions incorporated in grading rules from around the world. However, the recommendations from selected references set out in Table 3.1 show that there is some diversity of opinion amongst workers dealing with different species, and this emphasises the difficulty of generalising about wood properties.

For some timbers the ratio of strength properties parallel and perpendicular to the grain is relatively small. For example, tests of *Eucalyptus diversicolor* by Langlands (1936) and of *E. marginata* by Kloot and Schuster (1958) showed that strengths of these timbers were relatively insensitive to sloping grain. Ratios for modulus of rupture parallel and perpendicular to the grain were only 5:1 for *E. diversicolor* and 7:1 for *E. marginata*.

Kloot and Schuster (1958) went on to consider what reduction factors were appropriate to *E. marginata* with sloping grain when used as a building timber. Their methods and rationale are worth describing in full, both for their scientific content and their appreciation of practical issues. First, they rejected total reliance on the mean trend line for strength on grain angle because, for practical purposes, it takes no account of the variability encountered. They therefore prepared three trend lines for modulus of rupture on slope of grain (Fig. 3.4). The upper line is

Table 3.2. Reduction factors for the effect of cross-grain on the bending properties of *Eucalyptus marginata*. (Kloot and Schuster 1958)

Slope of grain	Reduction factors	
	Modulus of rupture	Modulus of elasticity
Straight	1.00	1.00
1:20	0.92	0.91
1:18	0.91	0.90
1:16	0.90	0.89
1:14	0.88	0.87
1:12	0.86	0.85
1:10	0.84	0.81
1:8	0.80	0.77
1:6	0.73	0.68
1:4	0.60	0.53
1:3	0.48	0.37

that for the mean of all values. The second line indicates values below which not more than 1 in 20 of the pieces would fall for any given grain angle, and the third line is constructed so that not more than 1 in 100 would have strength values below it.

Kloot and Schuster emphasised that the choice of the 95% line or the 99% line, or any other lower limit, is a matter of judgement, but it is essential to fully understand their significance. The 95% line, for example, indicates a reduction in strength of 23% for a 1:6 slope of grain. This means that only one piece in every 20 with this slope of grain will have bending strength less than 77% of what it would have had if the grain were straight. However, if 1:6 were chosen as the limiting slope of grain for sorting a large parcel of timber, all slopes from straight to 1:6 would be accepted. Therefore, the probability that any one piece picked would have a strength lower than that predicted by the regression might be as little as 1 in 100 or more.

Consideration is then given to the practical aspects of permitting larger grain angles in building timber for *Eucalyptus marginata*. The risks and consequences of underestimating grain angle, together with the probability that seasoning checks would develop more severely in cross-grained timber, thus further reducing strength, led to the conclusion that the 99% line would be the appropriate choice. This would lead to the reduction factors in Table 3.2.

Considering working stresses designed to produce a 72.5 and 55.5% stress grade, the grading rules might accept sloping grain of 1:6 for the former and 1:4 for the latter, if cross-grain were the only major defect. Three reasons are given for rejecting the acceptance of such timber for practical purposes. It would warp severely; it would develop severe seasoning checks; and the majority of pieces would have a much lower grain angle so that they would be used inefficiently, because higher working stresses could be justified for them. An alternative to liberalising the grading rules so extensively would be to set some realistic limit to grain angle, and to increase the working stresses permitted for this particular timber.

This example has been set out in some detail because it illustrates three important points. Uniform grading rules with respect to permissible grain angles cannot be applied to all timbers. Secondly, liberalisation of acceptable grain angles can lead to practical problems in other respects (warping, checking, and so on). Thirdly, it may be better to establish working stresses appropriate to the actual strength reduction from grain angles that are acceptable in practice.

With respect to this final point about determining appropriate working stresses, it may not be necessary to carry out detailed tests for each strength property, for each species, at a wide range of grain angles. Application of the following formula, known as Hankinson's formula, involving standard strength properties measured in straight-grained material, would seem to provide satisfactory values:

$$Sx = \frac{S_{\parallel} \cdot S_{\perp}}{S_{\parallel} \cdot \sin^n x + S_{\perp} \cdot \cos^n x},$$

where

Sx = strength of wood in which the grain is inclined at an angle x° to the direction of the load;

$S_{\parallel}$ = strength parallel to the grain (x = 0°);

$S_{\perp}$ = strength perpendicular to the grain (x = 90°); and

n = a constant for the particular strength property.

Kollmann and Côté (1968) suggested that the following values of n are appropriate:

Tension:	$n = 1.5-2.0$
Compression:	$n = 2.5$
Bending (MOR):	$n = 2.0$
Toughness:	$n = 1.5-2.0$

The denser, and the more homogeneous the wood, the smaller the differences are likely to be between strength measured parallel or perpendicular to the grain (Kollmann and Côté 1968). Oak and lignum vitae, for example, are known to have low strength ratios in the order of 1.1 to 3.4. Such timbers justify special consideration when applying corrections to strength properties for grain angle.

Northcott (1965) cast some doubt on the general applicability of Hankinson's formula to a wide range of timber species in that it ignores the shear strength for a species, so it should perhaps not be taken for granted under all circumstances. However, Langlands (1933) in tests with *Eucalyptus diversicolor* (a hardwood with small anisotropy) and Kingston (1947) in testing *Araucaria cunninghamii* (a softwood with pronounced anisotropy) both found that Hankinson's formula gave close agreement with experimental results.

A paper by Langlands (1936) contains a useful summary of the history of measuring the strength of wood with cross-grain, and of the various formulae that have been proposed. As a result of this review Langlands maintains that Hankinson's formula provides closest agreement with tests on a wide range of species that include *Acer* sp., *Fraxinus* sp., *Juglans* sp., *Picea* spp., *Pinus elliottii*, *P. strobus*, *P. sylvestris*, *Pseudotsuga menziesii*, and *Swietenia* sp.

3.3.2 Strength of Tree Trunks and Roundwood

Thunell (1951) carried out a series of tests that led him to suggest that spirally grained trees are better able to meet certain environmental extremes. His first experiment used models in the form of spirally wound and axial cylinders of identical dimensions. These were held at one end and loaded at the other to examine the response to both bending and torsion. The spirally wound cylinder showed greater deformation under the bending load, but less when tested in torsion.

A second series of tests was then made using tree stems of similar dimensions, and similar external appearance apart from the fact that some were straight-grained and some had severe spiral grain. These were first tested below the elastic limit in torsion using a range of turning moments. Then the stems were supported at the ends and bent, using central loading, up to breaking point, whilst measuring deflections under increasing load. Finally, the unbroken ends were reduced in length to four times the diameter, and tested in compression parallel to the axis.

The stems with spiral grain showed lower deformation in torsion and an initially higher deformation in bending, but the straight-grained stems deformed more under higher loads. In compression, the initial linear part of the deformation curve was much greater in spirally grained stems, but eventual breaking was by splitting, whereas straight-grained stems failed by compression of the fibres.

On the basis of these results Thunell (1951) inferred that trees with spiral grain are better equipped than straight-grained trees to withstand high winds and heavy loads of wet snow. Maximum wind resistance is reduced by bending of the stem and branches, and as they bend the branches will also shed snow more readily.

Champion (1929) cited Kindseth, a Norwegian forestry officer, as saying that spiral grain is essential for tree survival in the high mountains because it makes the trees more resistant to snow and wind damage. He even stressed the point to the extent of recommending that at the upper timber limit, where forest cover is more important than timber production, seed from trees with spiral grain should be sown by preference. However, this view was not held by Venkataramanan (1967) who maintained that *Pinus roxburghii* with spiral grain is especially liable to snow break and damage in felling as it lacks strength to withstand large tension and bending strains.

The strength of posts and poles with spiral grain has caused less concern than their tendency to twist with changing moisture content. However, Sanyal et al. (1980) found that the bending strength of *Pinus roxburghii* poles was unaffected by grain angles up to 12°, though considerable strength reduction was observed with higher grain angles. They also made the point that spirally grained poles fail progressively by splitting and partial delamination of growth layers, which gives more warning than the splintering tension failures common in straight-grained poles.

This tendency to give warning of failure, together with the extensive linear deformation with load before failure in compression as observed by Thunell (1951), would appear to be favourable features for pit-props and mining timbers. König (1947) specifically referred to the use of posts with moderate spiral grain as pit-props, i.e., as columns, but not for timbers withstanding lateral loads or tension. Nor should half-rounds be used if they have spiral grain.

3.4 Effect on Other Aspects of Wood Physics

In order to predict the non-steady-state concentration of moisture in wood it is necessary to determine the moisture conductivity coefficient, k. Wengert and Skaar (1978) examined the effect of spiral grain on this coefficient because they were aware that moisture can move 100 times more easily in the longitudinal than in the transverse directions (below 100 °C). Thus small grain angles would seem likely to have a large effect on the rate of moisture movement, and on the measured values of the conductivity coefficients.

For grain angle x°, the conductivity coefficient k_x can be calculated in terms of the conductivity coefficients in the longitudinal (k_L) and transverse (k_T) directions:

$$\frac{k_x}{k_T} = \cos^2 x + \frac{k_L}{k_T} \sin^2 x \ .$$

Calculations based on this formula show that small grain angles can indeed result in a large error in the measured value of k, e.g., 6° grain angle can result in 100% error where $k_L/k_T = 100$.

Although moisture will flow by circuitous paths, alternating between longitudinal and transverse flow, an overall effect of grain angle can be observed by examining $(F_L \sin x)/Fx$, the flux that moves in the x-direction by a longitudinal path, compared with the total net flux (Table 3.3) where:

$$(F_L \ \sin x)/Fx = \frac{1}{(k_T/k_L)\cot^2 x + 1} \ .$$

Wengert and Skaar (1978) made the point that this grain angle effect and the equations given can, by analogy, be applied to permeability measurements (fluid flow under pressure), and possibly to high temperature drying, where the ratio of longitudinal to transverse permeability is in the order of 20000:1; to electrical

Table 3.3. The fraction F_L/F_x of the total flux in the x direction which moves by the longitudinal path for various values of grain angle x when $k_L/k_T = 10$, 50, and 100. (Wengert and Skaar 1978)

x (°)	k_L/k_T 10	50	100
0	0.00	0.00	0.00
2	0.01	0.06	0.11
4	0.05	0.20	0.33
6	0.10	0.36	0.52
8	0.16	0.50	0.66
10	0.24	0.61	0.76
20	0.57	0.87	0.93
40	0.84	0.97	0.99
60	0.97	0.99	1.00
90	1.00	1.00	1.00

Table 3.4. The fraction of the total flux in the x direction which moves by a longitudinal path for various angles of x, for electrical, thermal, and sound conductivity ($k_L/k_T = 2$ to 5), and permeability ($k_L/k_T = 20000$). (Wengert and Skaar 1978)

x (°)	Electrical, thermal, sound	Permeability
0	0.00	0.00
5	0.02	0.99
10	0.09	1.00
20	0.28	1.00
45	0.75	1.00
90	1.00	1.00

and thermal conductivity measurements where the ratio is approximately 2:1 to 3:1; and to sound velocity measurements where the ratio is approximately 3:1 to 5:1. Table 3.4 shows the ratio F_L/F_x for these examples.

Archer (1979) examined growth stress distributions in trees, using the hypothesis that longitudinal and circumferential growth strains are continuously induced at the periphery of the growing stem. A plane strain combined with a pure torsion model was used to compute the internal stresses and strains due to forces and moments caused by the new growth increment. The twisting angle of the pure torsion model is caused by the shear stresses set up in the growth increment as the growth strains are induced along the grain axis, and the coupling of axial and torsional effects due to the elastic constants for the inclined grain material. Detailed stress distributions were derived for a range of cases in each of which grain angle was constant. Archer found that torsional shear distribution was extremely sensitive to quite small grain angles.

3.5 Effect on Conversion, Use and Economics

The effects of spiral grain on strength and stability of sawn timber, plywood, and roundwood have already been dealt with in earlier sections of this chapter because they constitute major defects in wood, and because their basic causes need to be set out in some detail. However, there are other aspects of log conversion and wood use that are also affected by spiral grain, and these will be considered at this point, together with the practical consequences of warping and reduced strength due to spirality.

3.5.1 Effect on Conversion

Mayer-Wegelin (1956) stated that splits in logs with spiral grain do more harm than in straight-grained logs because the defect crosses more boards, and Northcott (1959) described the combination of star shake and spiral grain as one of the most serious defects for the manufacturer of sawn timber or plywood. A star shake comprises a number of splits radiating from the pith. Occasionally this type of defect can be induced by bad felling practice, but as the shakes commonly contain considerable deposits of aged (gummy or crystalline) resin, it is apparent that they frequently develop in the standing tree. The splits that comprise a star shake naturally run with the grain along the stem, and in a spirally grained tree they therefore run around the stem. The consequence is that a defect that would appear as a longitudinal streak in boards or veneer cut from a straight-grained tree shows up as a diagonal defect cutting across the wood from a spirally grained tree.

A spiral angle of 1° will induce a change of 48° in the angle of a heart shake at the opposite ends of a 5-m log (McBride 1967). Similarly, a 2° spiral results in a 96° twist. Thus a single heart shake, when associated with a small spiral angle that would have no effect on timber grade per se, will degrade much of the timber cut from the centre of a log.

McBride (1967) described a study of 33 logs of *Pseudotsuga menziesii* sawn predominantly into 50-mm dimension of the type used for light construction. In this study it was assumed that a split in a straight-grained tree could be confined to a single board. The pricing of the out-turn was therefore first based on actual grades as sawn, and then this value was compared to the grade as if each defect were confined to a single piece. The weighted average spiral index (Brazier 1965) was only 4.6° for these logs, but the value was reduced by 26% due to the combination of shake and spiral grain. In a similar group of 19 logs with no heart shake and with spiral grain index of 2.8° no loss in value was recorded for spiral grain. In four logs of *Picea engelmannii* with heart shakes and a spiral grain index of 5.3°, the loss in values was estimated at 13%.

On the British Columbia coast, logs are scaled and graded by rules that divide them into value classes. Spiral grain is recognised as a defect which reduces log values. McBride (1967) described a study by the British Columbia Forest Service of 5730 logs of *Pseudotsuga menziesii* and 4745 logs of *Tsuga heterophylla*. Of the former, 7.4% were degraded for spiral grain, reducing the value by 4.2%, and of the latter 1.6% were degraded, reducing the value by 0.9%. On the basis of the total cut of logs in British Columbia, McBride estimated that, according to 1966 prices, spiral grain resulted in an annual reduction in value of $ 5 million for these two species.

These logs were cut from virgin forests, and some of the extreme spiral angles would have developed in the outer wood of senescent trees. In this respect there is no way to reduce the frequency of spiral grain, but the very magnitude of the cost to production suggests that other studies should be undertaken to examine the true costs of spiral grain at all stages of production and use in species where it is known to be present.

A study of 529 trees of *Fagus sylvatica* from five areas was described by Krahl-Urban (1967). Of these, 129 trees (26%) were free of spiral grain, and of the 400 trees with spiral grain 265 (66%) had LH spiral, 132 (33%) had RH spiral, and 3 (1%) minor grain fluctuations. In general, grain deviations of 7 cm m^{-1} (1 : 14) or more were the basis of significant reduction in value, and although the average spirality for the five logging areas was in the narrow range 4.4 to 4.9 cm m^{-1} there were numerous stems in which grain angle exceeded 7 cm m^{-1}. The sawn outturn was graded both with reference to grain angle and without it as shown

Table 3.5. Grading of timber sawn from 529 *Fagus sylvatica*. (Krahl-Urban 1967)

Timber grade	Including assessment of grain angle		Disregarding grain angle	
	(m^3)	(%)	(m^3)	(%)
A	144.30	13	211.27	18
B	728.21	61	718.48	59
C	321.28	26	275.43	23
Total	1193.79	100	1205.18	100

in Table 3.5. Krahl-Urban estimated that spiral grain reduced the total value by 7%.

In many plantation-grown softwood timbers, it is the initial spiral in wood surrounding the pith that should be segregated as far as possible by appropriate sawing patterns. In this way low-density corewood, which is so prone to twist, can be sawn to small dimensions in which warping can be more easily restrained during drying (Banks 1969, de Villiers 1973).

3.5.2 Effects on Wood Use

Efficient drying under restraint will minimise the problems of twisting warp arising from spiral grain, if the timber is dried to approximately the moisture content it will encounter in final service. The need for restraint can be seen in Fig. 3.5 in which *Pinus radiata* with spiral grain has twisted so much in the one direction that the whole stack has been tilted. In extreme cases kiln-dried loads of spirally grained *P. roxburghii* have been known to press against the sides of a kiln causing damage to walls and doors (Banks 1969).

In practice, any shipment of timber with spiral grain that remains completely free of twist tends, unfortunately, to be the exception. Variability giving rise to extremes of moisture content within a kiln load after drying will cause timber to twist as it comes into equilibrium with ambient conditions, as will the transfer of dried timber to an environment with different e.m.c.

McBride (1967) cited a study of seasoning degrade in *Tsuga heterophylla* where the loss in value of the clear grades was \$ 2.92 m^{-3} during drying and planing. Of this, \$ 0.47 m^{-3} was attributable to warp and \$ 0.21 m^{-3} to torn grain. Unfortunately, the causes of the warp and torn grain were not determined, but McBride suggested that if one-half were attributable to spiral grain, the loss would still be \$ 0.34 m^{-3}. In common grades, where the drying was to 13–18% m.c. (as compared with 10% for clear grades) losses due to warp were only \$ 0.27 m^{-3}. McBride noted that clear boards often come from the outside of logs, where spiral grain is most severe, which may contribute to the higher incidence of warp in better grades.

Difficulties in machining wood with spiral grain arise from differences in grain direction on opposite sides of a board as explained in Section 3.2.1. The grain

Fig. 3.5. Kiln drying of spiral-grained *Pinus radiata*. Twist in individual boards is all in the same direction and has caused the whole stack to twist. (Photo J. A. Kininmonth)

tends to tear or pluck out on the side where cutting is "against the fibre", i.e., where the grain angle goes downward from the face of the board in relation to the direction of planing. Brazier (1965) found that the amount of tearing in *Larix × eurolepis* due to spiral grain was correlated with the spiral grain index ($r = 0.65$), significant at the 99% level of probability.

Apart from tearing of the wood fibres, a severely twisted board, particularly if flat-sawn, is inclined to split under the pressure of the rollers, and cause jamming of the side cutters in a four-sider, which may result in damage of a serious nature (Rault and Marsh 1952). Splits in timber with spiral grain also cause problems in sawing and tend to jam the saw blade (Thunell 1951).

The problem of trying to straighten and fasten twisted boards and scantlings is one that every wood user has had to face from time to time. Simple twist in moderate degree is probably easier to rectify in a board than other forms of warping such as crook, but in extreme cases the force required to eliminate twist can cause breakage (Sears 1950). Yamamoto et al. (1978) suggested that timber of *Larix kaempferi*, which distorts severely in drying, should be used green for framing so that mutual restraint within the structure can be used to reduce distortion during natural drying. In New Zealand, house framing with green *Pinus radiata* is common practice, but it has been found necessary to incorporate in building standards a requirement that the frame should have dried out before enclosure!

Mention has already been made of the preference for straight-grained logs, or logs with consistent spiral direction, for split products such as shingles and cooperage (Burger 1941). For items of high value, and for which straight grain is essential, such as impact handles, splitting into blanks before turning has been recommended despite the wastage of wood involved compared with sawing (Langlands 1932). The North American Indians are said to have taken equal care in making their canoe paddles (Sears 1950). However, the most expensive example of splitting before sawing was that employed to ensure that World War I aircraft wing beams of *Picea sitchensis* were absolutely straight-grained. Logs were first split into quarters before sawing as closely as possible to the indicated grain direction (Tiemann 1941).

When wood is used for bending into curved elements for furniture and so on, straight grain is essential. König (1947) stated that grain must not exceed 3 cm in 1 m (1:33) in *Fagus sylvatica* when used for this purpose. On the other hand, König explained that spiral grain in *Fagus*, as in other hardwoods, does not rule out uses where short lengths can be utilised. In particular, turnery can make use of short pieces, and rotation of the work in this process ensures a fine finish irrespective of grain direction.

The actual costs of spiral grain in terms of wood wastage, rejection of unsuitable material, and failures during manufacture are difficult to assess. At times it may be very high. Champion (1924) stated that in some areas of India 50% of the growing stock of *Pinus roxburghii* had to be rejected, even for the relatively non-critical use as railway sleepers, on account of spiral grain. Lowery (1967a) attempted to predict the effect of spiral grain on the value of *Pseudotsuga menziesii* recovered in four mill studies. He used ten independent variables based on various aspects of spirality, but concluded that the coefficient of correlation ($r = 0.69$) based on the eight best variables had poor predictive value, accounting for less than half of the losses.

Apart from the papers by McBride (1967) and Krahl-Urban (1967) on economic losses from spiral grain, little other information is available. McBride made a plea for further studies including (1) the evaluation of losses in the plywood industry through warped plywood and rejected veneers, (2) assessment of the volume of construction timber degraded due to spiral grain, (3) assessment of the costs of replacing twisted sections, of damage to structures and manufactured articles, and of reduced acceptance of wood. So far no one seems to have taken up the challenge.

In the absence of any major effect on wood anatomy, it seems logical that wood with spiral grain will not prove unsuitable for pulping. Bhat and Singh (1955) found that *Pinus roxburghii* with spiral grain was satisfactory for manufacture of sulphate pulp for wrapping papers, and this was confirmed in a commerical scale trial at a paper mill in North India.

Finally, mention should be made of one use that found severe spiral grain to be advantageous. Edwards (1931) reported that in the 18th and 19th centuries trees with LH spiral grain were sought out for the curved mold-board section of the so-called wooden plough of the period. The length of the mold-board was determined to some extent by the angle of the grain because cutting across the grain had to be kept to a minimum to preserve the strength of the curved surface. For this reason a large hardwood tree with pronounced LH spiral was greatly prized by the plough maker.

3.6 Effect of Interlocked Grain

A few timbers with interlocked grain are greatly sought after because the striped figure produced in quarter-sawn or sliced material can be used to great advantage in furniture and joinery. However, interlocked grain is also a feature of many timbers that are unsuitable for furniture, but which are used in a great variety of ways, from plywood to heavy construction. In these situations interlocked grain may be quite undesirable because of its effects on wood strength, shrinkage, and machining properties. A possible exception is provided by the pattern of surface checking that arises in flat-sawn timbers with interlocked grain. This appears to resist wear far better than quarter-sawn timber or timber with only irregular grain deviations (Fig. 3.6). It is as if the potential "loose ends" remain "tucked in" and effectively hold the timber shape.

When interlocked grain is required for its decorative value, it is important that a log buyer should be able to evaluate it before conversion. Scheiber (1973) described a number of ways to evaluate intensity (angular contrast between changes of grain direction) and the frequency of changes in angle across the radius. Timber with basic density in the range of 350–500 kg m^{-3} often shows rings on the cross-section corresponding to the way fibres, in contrasting directions, are raised by the saw cut. Zigzag end checks may also provide a good indication of both frequency and intensity. Other methods involve taking some sort of sample such as a thin disc, or by using a increment borer.

Interlocked grain naturally poses problems in planing because, no matter which direction is used, alternate "stripes" are always facing the wrong way for a smooth finish. Scheiber (1973) suggested that best results can be obtained by reducing the cutting angle and increasing the sharpness angle compared with those used for straight-grained timber.

The twisting that arises when wood with simple spiral grain is dried, is complicated by the alternating directions of spirality in interlocked grain. In practice, grain angles may differ to left and to right, and may also be of different duration, i.e., occupying different proportions of the radius. These variations produce complex patterns of distortion in drying and make prediction impossible. However, Takemure (1978) provided some insight into how laminae with different grain angles interact with changing moisture content. He examined interlocked grain in a flat-sawn board, treating it as a composite material in which orthotropic plates with constant angle between their constituent fibres are laminated together. Torsional distortion depends on stresses between the plates during drying, the shear modulus of the plates, and the thickness of the board. An interesting conclusion was that the angle of torsion per unit length reaches a maximum when the angle of inclination of the fibres is approximately 10° or 80°, and a minimum at 45°.

The most obvious effect of interlocked grain on mechanical properties of wood is that it makes it difficult to split. This may be an advantage for heavy uses such as driveway planking and for large rollers like those used in house moving (Koehler 1949). Langlands (1933) also found that toughness was greater in specimens of *Eucalyptus diversicolor* with interlocked grain than in those with straight grain. However, Weddell (1960) studied bending strength in *Entandrophragma utile* and *Ocotea rodiaei* and found that both ultimate bending stress and

Fig. 3.6. Well-weathered wharf decking. From *left to right*: (1) *Eucalyptus* sp. flat-sawn with slightly wavy grain, (steel rail), (2) moderate but changeable grain angles, (3) complex interlocked grain, flat-sawn, (4) interlocked grain, quarter-sawn (5) relatively straight grain and (6) (*extreme right*) treated *Pinus* sp. with straight grain between knot whorls. (Photo H. G. Hiemcke-Hemming)

modulus of elasticity were adversely affected by the presence of interlocked grain, the effect on ultimate bending stress being particularly marked. Loss of strength was less severe and failure occurred less suddenly when beams were loaded on the tangential face than on the radial face. The effect of interlocked grain was, however, less than the effect of simple spiral grain. Nevertheless, Weddell recommended that pieces with severe interlocked grain should not be used where bending strength is important.

Weddell's results are contrary to the expectations of some writers (e.g., Kadambi 1951) who believed that interlocked grain would make for stronger beams than straight grain. This impression may be due to the fact that many hardwoods, well known for their interlocked grain, are intrinsically dense, strong timbers, and because the "localised" grain deviations are compensated by bands running in the opposite direction.

As with shrinkage of interlocked grain, the effect of strength may depend on the angular contrast between successive spiral bands, the degree of anisotropy in strength properties, and the inherent strength properties for a species. It may therefore be premature to suggest, as Weddell did, that other species with interlocked grain might be expected to give similar results to those that he studied.

3.7 Effect of Wavy Grain

Because wavy grain tends to be of sporadic occurrence, and is prized mainly for its decorative appearance, any difficulties in machining, finishing, and gluing are usually accepted as part of the price to be paid for achieving a special effect. Moreover, because wavy-grained wood is used mainly as boards or veneers in decorative situations, the effect of the grain deviations on strength is seldom an issue. However, Vintoniv (1981) examined some physical and mechanical properties of *Acer pseudoplatanus* with fine regular ripples, ridges 7–8 mm apart, and troughs 1.5–2 mm deep. Compressive strength parallel to the grain, modulus of elasticity in static bending, and wood density were all somewhat inferior to the properties of straight-grained wood, and longitudinal shrinkage was high.

3.8 Conclusions

1. Spiral grain is an insidious defect because it often goes undetected in sawn timber and veneers.
2. Strength of timber with spiral grain can be severely reduced, and shrinkage problems may arise, particularly twist in drying sawn timber and in plywood.
3. Twist in transmission poles with spiral grain can severely stress and even break transmission lines.
4. The economic cost of spiral grain for conversion and use of timber deserves much more research so that the true situation can be properly assessed. The few studies that have been conducted so far indicate that economic losses are considerable.

Chapter 4

Spiral Grain in Relation to the Environment

4.1 Effect of Environmental Variables

From earliest times it has been assumed that environmental factors could either induce spiral grain or influence its development. Meyer (1949) quoted Theophrastus of Erosos (372–287 B.C.) as saying that spirally grained trees *twisted by the north wind* were unsuitable for cutting. At different times wind, sun, soil, exposure, altitude, and many other features have been held responsible for "twisted" trees.

Insofar as most features of the environment will effectively operate in one "direction" throughout the life of a tree, it is apparent that they cannot be held responsible for all types of spirality, especially the changing spiral directions encountered in many species (Champion 1924, Meyer 1949, Mayer-Wegelin 1956, Nicholls 1965). On the other hand, it is almost axiomatic that any feature capable of modifying growth and development of a tree might be capable of influencing the intensity of spiral grain within it, whatever its origins (Harris 1981). For this reason, as well as due to historical interest, it is worth recording the various views that have been expressed, and theories that have been put forward over the years.

Many environmental features are inevitably linked together. Thus, wind, aspect, exposure, and altitude may be to some extent interdependent. Nevertheless, different arguments have been advanced in support of each environmental variable, and so they will be considered separately, even though this sometimes complicates attribution of the preferred cause to a particular author.

4.1.1 Movement of Earth and Sun

One of the most fanciful explanations put forward to explain spiral grain was attributed by Thunell (1951) and Krahl-Urban (1953b) to A.H. Kennedy. The rotational velocity of the earth and its oblate shape are supposed to create spiral and vortex movements of the air around trees, because the southern sides of trees in the Northern Hemisphere and the northern sides in the Southern Hemisphere are subject to greater velocities than the corresponding opposite sides. The air currents so produced are supposed to operate via the branches to twist the stem.

Van Oye (1926) considered the effects of roots, branches, sunlight, soil, and slope, and also came to the conclusion that only the rotation of the earth could account for spiral grain, which he believed to be predominantly in one direction, from left to right. He could only assume that the earth's rotation in some way affected the roots, which, if developed asymmetrically, would induce torsion in the stem. It is of interest that this theory was quickly criticised by Champion (1927b) on the grounds that both LH and RH spiral grain had been observed (on both sides of the equator), and because of the changing spiral direction that he and other writers had observed in many conifers. However, Champion did express some interest in Van Oye's theories concerning the influence of asymmetrical lateral roots on spiral grain, as will be described in Section 4.1.2.

Despite valid and lucid criticism against ascribing the initiation of spiral grain to any unidirectional force, such ideas continued to appear in print for many years. For example, Butler (1931) had observed RH spiral grain in apple trees, and had been told that trees in the Southern Hemisphere spiralled in the opposite direction. He therefore decided that the condition is a kind of heliotropism "similar to the turn of the sun-flower and the leaves of the compass plant of the western prairies".

Many writers have drawn attention to spiral growth as a recurrent feature at all levels of biological and physical development (Braun 1854, Champion 1924, Copisarov 1933, Seifriz 1933a, b, Preston 1949), and have drawn analogies with spiral growth in other plants. In a carefully reasoned paper, McKinney and Sando (1934) reported twisting in the rachis of certain varieties of wheat when grown under short-day conditions, and in certain other varieties only when grown under low light intensities. In either case the direction of twist was random. From this they deduced that the propensity to twist is controlled by heritable factors, but that its expression is dependent on environment. They therefore suggested that trees should be grown under different day lengths and light intensities to measure the response in spiral grain, and that spirality should be assessed from north to south of the range for tree species with widespread habitat, to determine the effects of sunlight at different latitudes.

4.1.2 Soil Conditions

As will become apparent, it has proved very difficult to generalise about environments that encourage the formation of spiral grain in trees, but there has been very broad acceptance of the idea that harsh growing conditions favour its development. Such conditions have often been taken to include sites with poor soils in moorlands or mountainous regions (e.g., Gunther 1929, Koehler 1931, Kadambi 1951, Thunell 1951, Kaasa 1976). Indeed, there is no need to rely on science for this observation. The windswept twisted tree of the "blasted heath" has also fascinated artists around the world, to the point where it has almost become a visual cliche to indicate a harsh environment.

The earliest reports on *Pinus roxburghii* in India suggested that the underlying rock had a considerable effect on spiral grain (Smythies 1915). Its prevalence on mica schists seemed to be particularly significant at first, but many exceptions were found later, and Champion (1924) took a more cautious approach towards ascribing spiral grain to any particular soil type.

An explanation of the way in which nutrient status of the soil might encourage spiral grain has been attempted by few writers. Raunecker (1957) believed that the chemical composition of the soil affected the development of spiral grain in *Pinus sylvestris*. He noted that trees grew straight grained but often contained compression wood when grown on river sands to the east of the river Main, whereas on adjacent alkaline soils (shell-lime soils) trees were frequently spirally grained, but generally lacked compression wood. Wood grown on the alkaline soils also had an abnormally high starch content (14% of dry wood substance in the sapwood) which Raunecker believed to be the result of very low levels of readily available phosphate. In his view the high starch levels arose because the essential polymerisation process from monosaccharide to oligosaccharide to polysaccharide was being impeded by lack of phosphate. He concluded that such trees were unable to respond to environmental pressures in the ordinary way (by forming compression wood) but formed spiral grain instead.

Copisarov (1933), in an unusually florid paper, attempted to relate dendritic patterns formed by certain inorganic salts (the Liesegang phenomenon) with minerals involved in tree growth, to prove that "tree twisting is primarily due to local soil peculiarities which eventually modified the structural constituents of the plants. In the course of evolution the contact and inductive influences transformed the mineral efflorescence and arboreal growth into new structures more complex both in form and substance"!

McKinney and Sando (1934) were quick to indicate the fallacies in Copisarov's argument, and in fact neither this nor the paper by Raunecker have received much support. Later studies have found little influence of soil composition on spiral grain (e.g., Rault and Marsh 1952, Kadambi and Dabral 1955, Fielding 1967, Harris et al. 1979). However, it would be wrong to dismiss out of hand the possibility that lack of minerals essential for plant growth may in some way influence stem polarity.

Under field conditions it may sometimes prove difficult to identify primary causes. For example, Whyte et al. (1980) found that spiral grain in *Pinus caribaea* grown in Fiji was significantly less strongly developed on poor soils than on fertile soils – the comparison was between ferruginous and humic latosols. However, it was noted that treatment of the poorer soils with fertilisers did not increase spiral grain, and it was therefore concluded that the effect may have been associated with the better water-holding capacity of the humic latosols. Support for this was provided by the other statistically significant difference found between localities, in which higher spirality was found in areas of higher rainfall. Similar results were obtained by Smith et al. (1971) in their work with the closely related species *Pinus elliottii*, although Fielding (1967) had suggested that spirality in *Pinus radiata* was worse in drought years.

Mayer-Wegelin (1956) cited earlier reports in which both very dry and very wet soils had been said to favour spiral grain, a view that was also held by Kadambi (1951). On the other hand, partial droughting compared with irrigation, was not found to influence spiral grain in two, quite separate, controlled experiments with *Pinus radiata* (Nicholls and Waring 1977, Harris et al. 1979).

4.1.3 Roots

Soil characteristics will also affect root development, and several writers have discussed development of spiral grain in relation to root systems. Small (1932) described differences in stability of *Quercus virginiana* when grown on the open soils of shell middens compared with trees growing on undisturbed sands, and went on to consider the consequence of this in causing the trees to be twisted by the wind (Sect. 4.1.5).

Van Oye (1926) theorised that if lateral roots were not developed symmetrically about the base of a stem they would induce spiral grain, but that trees with symmetrical roots or strongly developed tap roots would not be so affected. When trees grow on shallow rocky soils the orientation of the roots may be dictated largely by cracks in the rocky substrate, and this has sometimes been related to the frequency of spiral grain in coastal and mountainous areas (Champion 1924, Gunther 1929, Cahn 1931, Mayer-Wegelin 1956, Venkataramanan 1967). However, no one has pursued this matter further and asked why putative induction of spirality in the base of a stem should lead to spiral development along its entire length. The evidence of weathered stumps points, in fact, to little more than a localised influence of root direction (e.g., Fig. 2.4).

Roots themselves develop spiral grain, sometimes quite intensively (Herrick 1932), which, incidentally, makes it difficult to envisage application of many of the exterior forces, such as wind, which have been suggested as being responsible for spirality in the stem.

The nature of the insertion of roots into the stem does have considerable significance for the development of the stem as a beam of uniform resistance to wind pressures on the crown (Ylinen 1953). In this respect the whole structure of the stem is dependent on, and may well respond to, the support and stabilising effect of the root system, as well as to the inputs of water and mineral nutrients from it.

4.1.4 Slope and Aspect

Smythies (1915) believed that *Pinus roxburghii* growing in India on southern and western aspects was more spirally grained than trees growing on northern and eastern aspects. However, Canning (1915) found no support for this in his area of India, and Champion (1924) decided that "a non-committal attitude is preferable". In a later study of the same species growing in South Africa, Rault and Marsh (1952) could find no relationship between aspect and spiral grain.

Van Oye (1926) believed that conditions on a slope were less favourable to the development of spiral grain than when trees were grown on the flat because he claimed that lateral roots tend to be produced mainly along the contours and cancel each other out. He then produced a suggestion, which greatly amused Champion (1927b), that to keep down the percentage of spiral grain, trees should be raised on the slopes and transplanted later!

4.1.5 Wind

For a period in the early 1930s spiral grain became something of a vogue as a topic for correspondence in the scientific journals of North America. Because of the widely acknowledged frequency of trees with severe spiral grain in exposed situations, wind was seen by many to be the most likely cause. Wanderers in the wild from Lake Kahnipiminanikok (which is in Ontario) to Kisatchie Wold (which is in Louisiana) were able to reflect on their vacation experiences, and (sometimes) to have their holiday snaps published (Cahn 1931, Herrick 1932).

The objections to wind as the primary cause of spiral grain are, of course, the same as for any other constant or regular feature. Nevertheless, even if wind is no more than a possible contributory source of so-called twisting, as Champion suggested in 1924, or if it does no more than initiate a twist which may be magnified by other means (e.g., Priestley 1945) it deserves some consideration.

However, as the chief protagonist wind was usually believed to twist entire tree trunks around their axes, and the mechanism required the crown of the tree to act, so to speak, as the handle of the corkscrew. The crown either developed heavier foliage to the south naturally (Wentworth 1931) or winter gales moulded it into a lopsided form extending downwind (Yeager 1931, Howard 1932). Prevailing winds from another direction at other times of the year hitting the side of crown then provided the torsion.

In opposition to this concept, Koehler (1931) pointed out that progressive twisting of a stem must surely create the largest spiral angles at the pith and lowest on the outside, whereas the reverse is more frequently the case. Herrick (1932) opposed the idea on the grounds that branches on the opposite sides of the stem should twist in opposite directions, and that, in any case, wind could hardly have induced the spiral grain that is frequently observed in roots. From distant China, Jacot (1931) observed *Thuja orientalis* and *Juniperus chinensis* on burial mounds, and found no evidence that their undoubted exposure had induced the variable spirality that he observed.

Breakage of leading shoot and branches by wind has also been suggested as a cause of spiral grain and this will be discussed in Section 4.1.9.

An experiment to examine the effect of torque applied mechanically to young stems was reported by Quirk et al. (1975). *Pinus resinosa* trees, 10 years of age, were twisted continuously either to the right or to the left, throughout an entire growing season. One internode (the one containing five annual growth layers) was held firmly at its base and torque was applied through its upper end and through the node above (i.e., through the top and bottom of the 4-year internode). At the commencement of the experiment grain angles on the outside of the internodes under test averaged about 5° LH.

Initially, trees were twisted by 20°, and twist was then increased periodically over 3 months to 180° and held for 2 months. Residual twist was measured 1 week after removing torque. Under LH torque, grain angle increased down the entire internode, but was most noticeable at the top. Under RH torque, grain angle was reversed only at the top, and lower down LH spiral was reduced only slightly, an indication that natural grain direction can be reversed only with difficulty. There were also some notable side effects: radial growth was enhanced by as much as 100%, but tracheid length decreased 20% as a result of the treatment.

Other experimental work to measure the effects of torsion on stem spirality was reported by Pyszyński (1977b). Hejnowicz (1973a) had observed that the RH spiral grain that predominates in *Aesculus hippocastanum* could not be related to anatomical events of the type that will be described in Chapter 5. He therefore

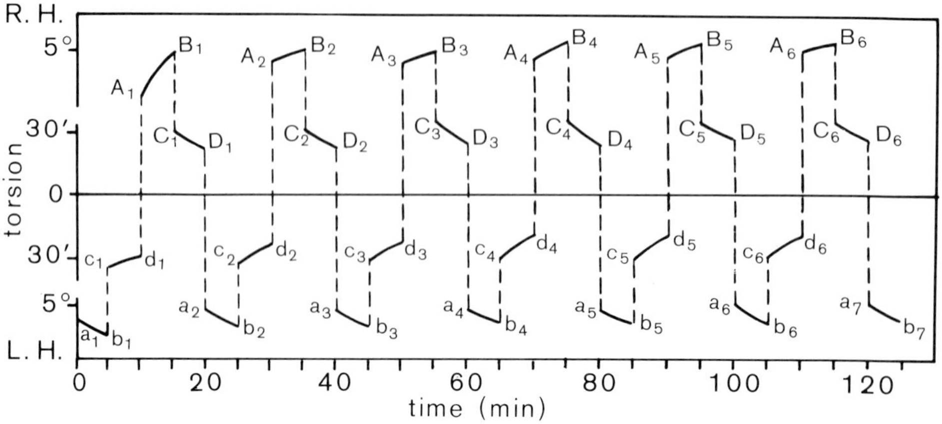

Fig. 4.1. Six cycles of torsion alternately to left (*small letters*) and to right (*capital letters*) applied to a young stem of *Aesculus hippocastanum* (After Pyszyński 1977b). Oa_1 = initial torsion to the left; a_1b_1 = further torsion over 5 min with the same load; b_1c_1 = initial recovery after load removal; c_1d_1 = further recovery over 5 min; d_1O = residual torsion

advanced a hypothesis of asymmetrical reversibility of torsion caused by wind. Because the microfibrillar helix in the major (S2) layer of the cell wall is RH, it was thought that relaxation of random torsion due to wind would be asymmetrical, and that irreversible RH twist would accumulate.

Pyszyński (1977b) took young stem sections of *Aesculus*, 800 mm long and varying in mean diameter from 14 to 38 mm. These were clamped at one end and subjected to cyclic torsion at the other (5 min in one direction followed by 5 min of relaxation, then 5 min in the other direction and a further 5 min of relaxation). This 20-min cycle was repeated several times. The result was that the stems became progressively more twisted to the right (or less strongly twisted to the left) after each period of relaxation, and it was concluded that random torsion would produce irreversible RH twist (Fig. 4.1).

Pyszyński (1977b) also presented evidence from standing trees of *Aesculus* in which branches had apparently been swung around as a result of the main trunk twisting to the right, so that the branches had pressed against neighbouring trees abrading their bark.

The evidence presented by Pyszyński (1977b) is convincing, but its applicability to the general problem of spiral grain is confusing. An RH helix is almost universal in the S2 layer of tracheid and fibre walls (Meylan and Butterfield 1978), and therefore all trees that grow straight grained initially should become progressively twisted to the right if the hypothesis of asymmetrical recovery from torsion is valid. It is difficult, for example, to understand why grain should remain straight in the wood adjacent to the pith in the majority of *Pinus* spp. if the (generally stronger) wood of *Aesculus* so readily becomes twisted.

4.1.6 Altitude

The broad associations of altitude with shallow rocky soils, exposure to wind, snow, and low temperatures are obvious enough, but there are other aspects that deserve consideration. Mayer-Wegelin (1956) suggested that some of the apparent effects of altitude on tree growth might be misleading. For example, high-country trees are both less accessible and (near the tree line, at least) of negligible commercial value. For these reasons they are allowed to grow to much greater age than trees in the managed forests of the lowlands, and therefore are more likely to develop the extreme spirality of senescence. This would certainly be true in Europe, but not necessarily so in North America, for example, where virgin forest is, or has been, available for comparison at all altitudes.

In India, trees growing at high altitude may miss some of the other pressures of human population, which include damage through grazing and fire, as well as the probability of preferential use of straight-grained trees by the villagers, leaving spirally grained trees behind as the breeding population (Smythies 1915, Mayer-Wegelin 1956), features that might favour the higher grain angles reported in the lowlands.

There are other instances in which grain angle has been reported to decrease with increasing altitude. Krahl-Urban (1953a, b) stated that stands of *Fagus sylvatica* growing at altitudes of 1000–1200 m in Switzerland and Bavaria were widely known to be free of spiral grain. Whether genetics or environment are responsible for this is not known, but Krahl-Urban speculated that the frequency of wet snow in these situations might preclude the growth of spirally grained trees (reasons not given).

Cown et al. (1983) surveyed the wood properties of *Pinus caribaea* in all the main plantation areas of Fiji and found that plantations over 300 m had relatively low mean grain angles (less than 3.4°), whereas in lowland plantations mean angles lay between 4.7° and 6.6°.

These exceptions apart, the view is widely held that many tree species growing at high altitude tend to develop extreme spirality, and that this may result, at least in part, from natural selection. That is, trees with spiral grain are believed to be better adapted to the harsh environment, and therefore more likely to survive and reproduce themselves (Champion 1929, Thunnel 1951).

This leads to a possibility which the writer has not seen discussed elsewhere, but which would seem to be a logical consequence of naturally selected populations arising in discrete areas. "Unselected" seeds transported by birds or wind from the lowlands to the mountains would be disadvantaged compared with the local population. On the other hand, there is no reason to believe that seeds transported down the slope would fail to compete successfully once they had the environment that would favour vigorous growth. Every forester will have seen gnarled and twisted trees growing in exposed situations that have produced vigorous progeny where seeds have landed in a sheltered, fertile place. Is it possible that exposed situations may sometimes provide the continuum and the seed source for production of the severely spiralled trees that seem to arise spontaneously amongst the relatively straight-grained majority in lowland forests?

4.1.7 Temperature

Mean temperature of the growing site has seldom been directly implicated in the development of spiral grain, but Kromhout and Toon (1977) found that spiral grain in *Pinus patula* was less in the cooler temperate regions of South Africa than in the warmer areas, particularly in outer wood more than 20 years old. The effect of altitude on spiral grain of *Pinus caribaea* grown in Fiji (Cown et al. 1983) might presumably be open to the same interpretation. These two studies can be used to illustrate the distinction that should be drawn between the effects of severe exposure at high altitude, and the effects of altitude as a possible modifying influence within a temperate environment. However, it may be noted that Rault and Marsh (1952) found no effect of temperature on spiral grain in *Pinus roxburghii* grown in South Africa.

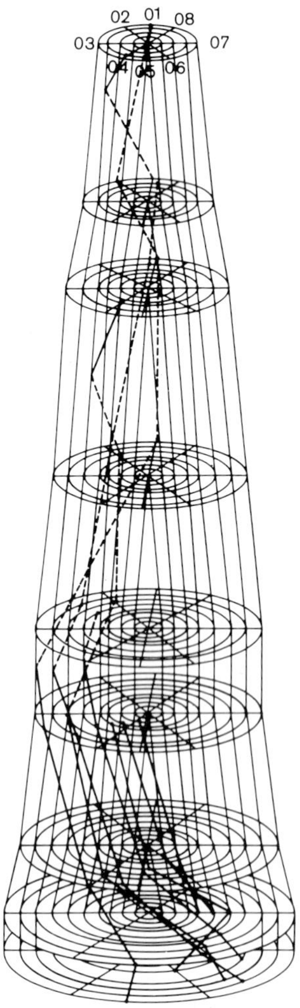

Fig. 4.2. Shift in the location of maximum width in successive radial increments in a sapling of *Pinus roxburghii* (After Misra 1943). At the front of the figure *solid lines* connect points of maximum width in each growth layer at different levels, and these are continued as *dotted lines* around the back of the stem

Extremely low temperatures, sufficient to damage trees and to cause cankers to form, were believed by Zalasky (1972) to be a factor in the formation of spiral grain in *Pinus banksiana, P. contorta, Populus balsamifera,* and *P. tremuloides.*

4.1.8 Eccentric Growth and Reaction Wood

Misra (1939, 1943) measured spiral angle, microfibril angle, and tracheid length in relation to stem eccentricity in *Pinus roxburghii.* He found that microfibril angle varied directly, and spiral angle inversely with ring width. He also found (1943) that the maximum radial increment in each annual growth layer followed a spiral course along the stem when measured at different heights. Furthermore, when measured on a transverse section, the locations of the maximum increments in successive annual growth layers also described the beginnings of a spiral (Fig.

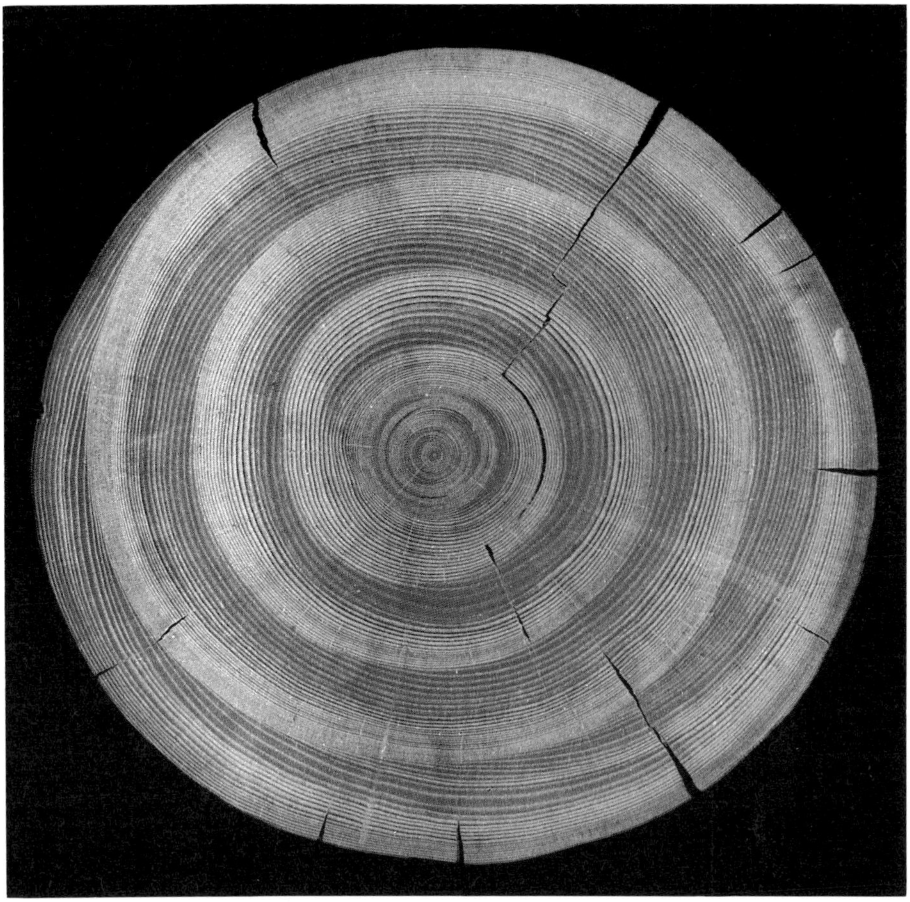

Fig. 4.3. Spiral compression wood in a cross-sectional disc of *Picea* sp. containing 137 growth rings. (Photo T. E. Timell)

Table 4.1. Relationship between stem lean and mean of four measurements of grain angle at 1.4 m. (Paterson 1967)

Degree of lean	No. of trees	Grain angles (°)	
		Mean	Range
Upright (0°)	44	4.1	0 – 11.0
Slight (1° – 3°)	45	3.7	0 – 9.5
Considerable (4° – 6°)	42	3.5	0.5 – 10.0
Severe (7° – 9°)	10	4.2	0 – 13.3
Very severe (10° +)	3	2.5	0.5 – 4.8

4.2). This curious phenomenon seems to bear some relationship to spiral compression wood (Timell 1986) in which the wide dark parts of successive growth layers describe a spiral in transverse section (Fig. 4.3).

The similarity between spiral compression wood and the spiral dye pattern in Fig. 1.2 suggests a similar interpretation. Let us suppose that the stimulus to form compression wood originates in the crown, and travels down the stem in a direction parallel to the wood grain of the outer growth layer. Then compression wood will form a spiral pattern in a transverse section if the spiral angle increases with stem age, and/or if the height of the stimulus source above ground level increases with increasing mean crown height.

Pawsey (1965) and Zobel (personal communication in Fielding 1967) have suggested that spiral grain in markedly eccentric logs abruptly becomes straight in that sector containing compression wood. However, Champion (1924) illustrated a tree of *Pinus roxburghii* with "pistol butt" (travelling horizontally at first before growing upright) in which spiral grain on the outside was nearly at right angles to the stem axis. Here, surely, is a situation with compression wood which has not obeyed the "no-spiral-grain" rule.

Pawsey himself (1965) found no obvious relationship between slope of grain and eccentricity of cross-sections of *Pinus radiata*, which led him to suggest that compression wood may not have been sufficiently developed in the mildly eccentric discs that he studied to have affected the slope of grain. Paterson (1967) also failed to discover any relationship between degree of lean in *Pinus patula* stems and the size of the grain angle, as shown in Table 4.1, which also illustrates the wide range of spirality encountered in each tree category.

4.1.9 Structural Breakage and Wounding

The concept that damage to trees is the fundamental reason for spiral grain can first be attributed to Smythies (1915), who associated it with burning, trampling, lopping, and grazing by local villagers and their flocks in the areas where *Pinus roxburghii* grows in India. Smythies believed that damage to young trees stimulated the development of spiral grain in unfavourable localities, and could even induce it in localities where it was otherwise less likely to develop.

In view of the fact that spiral grain is often evident even in plantations that have been well protected, Venkataramanan (1967) felt that injury was not necessary for the development of spirality, though he agreed that it could probably accentuate any inherent tendency. He referred to the likelihood that, under conditions of annual burning, a very large proportion of trees must be of coppice origin which was once thought to stimulate spiral grain, yet Troup (1916) and Champion (1924, 1927a, 1930) showed that coppicing does not appear to increase either its frequency or intensity. Kedharnath et al. (1976) similarly dismissed the commonly held belief that fire and injury are the main causes of wavy grain in *Pterocarpus santalinus*.

As a result of his observations of *Pinus contorta*, and to a lesser extent *P. flexilis* and *P. balfouriana*, growing at extreme altitudes in California, Gunther (1929) came to the conclusion that what he termed the "critical age", of a tree occurred when it first thrust its leading shoot above the general level of the canopy in its area. At this point the leader either became bent over or was snapped off by the ferocity of the wind at these altitudes (2500–3000 m). Gunther linked this type of damage, which inevitably affected every tree in such exposed conditions, with the onset of spiral grain, and in some way the further twisting of dead stems and branches as they dried out seemed to him to be a continuation of the same process, "age, or synonymously the presence of dead branches made to warp and twist under arid conditions, is actually responsible for the extraordinary angle of twist observed in dead branches and older trees".

The presence of trees without spiral grain in places where they would be expected, and the presence of undamaged trees with spiral grain in more sheltered areas, did not shake Gunther's faith in his hypothesis, though he conceded that "there seems to be no environmental law that holds good for the twisting of trees in general".

The effect of wounding is a topic that lends itself to the experimental approach, and several workers have deliberately damaged and wounded trees to see if such treatment would induce spiral grain.

Champion (1924) examined seedlings that had been damaged by various natural agencies, including snow and insect attack, and concluded that injuries do not generally result in spiral grain. However, he realised that his observations had been made over a short period, and whereas it was not reasonable to expect that wood well removed from the point of injury would immediately develop spirality, it would not be difficult to imagine a change spreading gradually outwards from an injury. He therefore commenced a long-term experiment. In 1921 he had tightly bound with wire 20 young trees of *Pinus roxburghii* in an area where spiral grain was severe, and 20 trees in an area with straight grain, but after 9 years (Champion 1930) he was unable to detect any difference in fibre direction above or below the constriction. A further 20 trees in each locality had also been partially girdled in 1921, but again, apart from callus formation and some grain distortion above the girdle, no general change in grain directon could be detected in 1931.

Bailey (1948) was interested in figured woods generally, including birdseye, burl, curly, wavy, and blister grain. He applied (with the ingenuity of a Torquemada) various forms of maltreatment (his own expression), including con-

Table 4.2. Persistence of needle traces and grain angle in *Pinus radiata*. (Harris 1969)

	Grain angle (°) at growth layer from the pith				
	1	2	3	4	5
A. Stems without needle traces beyond the 3rd growth layer					
	3	4	4	2	0
	5	3	3	2	1
	1	3	3	2	1
	3	3	3	3	3
	2	4	6	8	6
	5	5	5	3	1
	1	3	2	1	1
	3	4	4	4	3
	2	3	6	5	4
	3	5	8	7	5
	1	6	5	3	2
	2	4	7	6	6
	2	6	4	4	3
Av.	2.5	4.1	4.5	3.8	2.8
B. Stems with needle traces out to the 5th growth layer					
	1	2	2	3	3
	3	6	6	6	5
	2	6	8	8	6
	2	5	6	6	6
	6	5	4	5	5
	3	6	6	5	4
	3	7	9	9	9
	1	4	4	5	2
	1	5	8	8	6
Av.	2.4	5.1	5.9	6.1	5.1

striction, partial girdling, vertical cutting, bending, twisting, and pounding, to young trees of *Acer rubrum*, *Liriodendron tulipifera*, and *Betula aleghaniensis*, but although the stems (not unnaturally) became markedly disfigured, there was no evidence of the desired types of wood figure developing in any of them.

Harris (1969) suggested that the initial development of spiral grain in *Pinus radiata*, which normally increases from zero in wood near the pith to a maximum at about the third annual growth layer, is influenced by the spiral arrangement of the "needle traces", vascular supplies to the short shoots subtending needles. These persist for 3 years or more on the main stem, after which the short shoots and needles die and their vascular connections usually become occluded by normal cambial growth. However, in some trees occlusion is incomplete, so that bark intrusions continue and exaggerate the spiral pattern of the needle traces. Wood so affected tends to retain a wide spiral angle for longer than is normal. Similarly, stems that retain needles for more than 3 years also maintain large grain angles (Table 4.2).

Harris (1969) pursued this matter experimentally by inserting wooden pegs in the stem through bark, cambium, and immature xylem, at the scars left by the

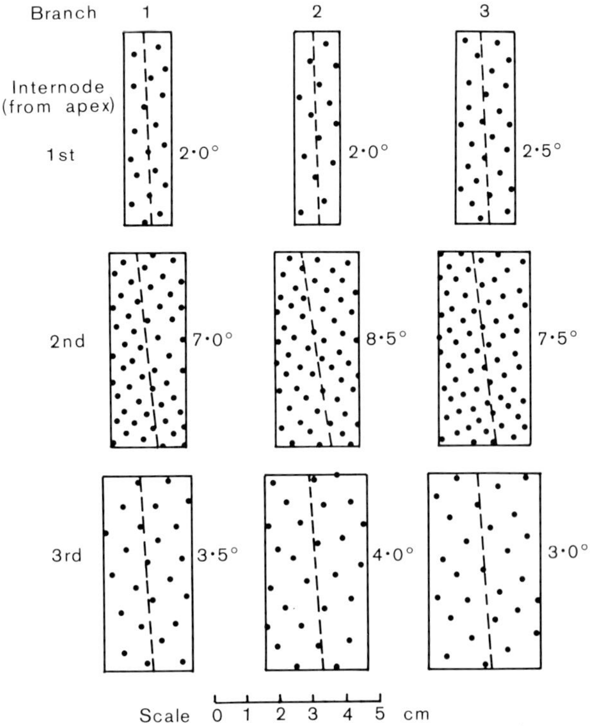

Fig. 4.4. Sections of branches from trees of *Pinus radiata* that had been artificially defoliated 2 years before felling. The diagram is drawn as if the bark had been slit longitudinally and laid out flat to show the parastichies of the short shoots. *Straight broken lines* indicate grain angle in the xylem of each section. (After Harris 1969)

short shoots on young stems in their fourth year of growth. This treatment was said to maintain the spiral angle in stems which elsewhere had smaller angles in the year's growth following loss of stem needles. In another experiment, all second- and third-year needles were plucked from young trees before the commencement of spring growth. Limited shoot elongation in the year after defoliation caused the parastichies, which had presumably already formed in the resting bud, to remain closely packed compared with growth in the preceding and following years. In Fig. 4.4 it can be seen how the parastichies on these branches would appear if the bark were split longitudinally and laid out flat. The grain angle in the underlying wood is also drawn in. While it is conceivable that the large grain angles in the condensed internodes could be attributed to a variety of causes, the implication is at least that branch extension reduced by defoliation has influenced spiral development.

These experiments have been presented in some detail because they not only show that (admittedly rather extreme and "unnatural") external factors can influence one particular component of spiral development, but they also suggest potentially variable structural features, the spacing, dimensions, and duration of needle traces, that may influence the degree of initial spirality in this species.

In this connection it may be noted that natural defoliation of *Pinus radiata* by the needle-cast fungus *Dothistroma pini*, which tends to infect mature needles predominantly from the base of the crown upwards, could not be shown to have any effect on spiral grain in the main stem (Harris and McConchie 1978). On the other hand, deliberate suppression of tree height growth ("hedging"), in which *Pinus radiata* trees kept to a height of 2 m were compared with normal trees, showed that the hedged trees had larger spiral grain angles (and reduced radial growth rate) up to age 13 years (Nicholls and Brown 1976). This difference, which was statistically significant only for rings 10 and 11, resulted mainly from a lower rate of decline from the maximum grain angles normally reached by ring 2.

4.1.10 The Case for Environmental Influence

Although there is considerable evidence that environmental extremes favour the development of spiral grain, controlled experimentation, and detailed observation have often failed to demonstrate any significant direct effect of the environmental variables being examined. Thus studies of geographical location (Lowery 1966, Woodfin 1969), site class (Krahl-Urban 1960, Krempl 1970), soil nutrients (Rault and Marsh 1952, Harris et al. 1979), and soil moisture (Nicholls and Waring 1977, Harris et al. 1979) have failed to prove any effect on spirality.

The role of genetics in determining spiral characteristics of different tree species will be dealt with separately (Chap. 6), but there is considerable support for the concept that the propensity for spiral grain is controlled by heritable factors, yet its expression is dependent, at least in part, on environment (e.g., Champion 1924, McKinney and Sando 1934, Richens 1945, Schmucker 1956).

Champion (1945) gave an example of *Pinus sylvestris* seed (presumably collected from straight trees in Germany) that produced badly spiralled trees when planted in southern France, outside its natural environment. Sears (1950) also stated that *Pinus sylvestris* seed from northern Europe produced trees of good form when introduced into North America, whereas reputable seed from southern Germany was said to have produced freak forms with grotesquely twisted stems and spiral grain. Although both of these references attach more trust than may be justified to the fidelity of commercial seed suppliers, most foresters could provide similar, if less extreme, examples of reputable seed producing inferior trees when grown in an environment different from that where the parent trees grew.

This raises the question as to whether many of the environmental extremes that are alleged to produce spiral grain are, in fact, not directly causative, but only (1) provide conditions that favour the expression of a spiral propensity already present genetically and/or (2) place constraints on the growth of trees without the propensity for spiral grain so that natural selection produces a predominantly spiralled crop. So far there seems to be no clear-cut indication of the true role of many environmental variables. Equally, however, there is sufficient evidence of environmental influence on many aspects of tree growth and development not to rule out the possibility of more direct cause and effect with respect to spiral grain.

4.2 Effect of Growth Rate

Rate of growth has always been a matter of concern to foresters, and, for present-day plantation forestry in particular, the economic consequences of high interest rates on capital investment have placed even greater emphasis on trying to shorten rotations through rapid growth. However, it is necessary to point out that although the rate of growth may provide a convenient yardstick for timber producers, it is often difficult to attribute to a single cause. For example, growth rate may be associated with age, radial growth near the pith usually being faster than further out in the stem, or it may be associated with a whole range of conditions at the growing site. Thus comparisons of spiral grain at different growth rates may involve circumstances much more complex than simply the rates at which cambial cells divide.

The effect of growth rate on spiral grain was reviewed by Sachsse (1965). He found that the experimental results available at that time were inconclusive, some indicating that larger spiral angles were associated with faster growth, some associating larger angles with slower growth, and some suggesting that growth rate and spirality were independent. More recent results are little more conclusive, yet it would be wrong to assume that such diversity is meaningless, and that growth rate never has any significant effect on spirality. The possible reasons for conflicting results will be discussed in Section 4.3 but first it is necessary to review the evidence that is available.

4.2.1 Slow Growth, Increased Spirality

Sachsse (1965) used the observations of Burger (1946) on *Malus domestica* and *Pyrus communis* as his first example of slow radial growth producing large spiral angles. Spiral angle increased with age in both species, and therefore, presumably, with decreasing growth rate. Many other species follow a similar pattern, but few studies have attempted to find out whether the effect is directly attributable to growth rate, or to other factors associated with tree age.

The question was addressed directly by Kennedy and Elliott (1957). They compared trees of *Alnus rubra* growing on a good growing site and a poor one. At comparable diameters, the trees on the poor site had much greater angles than those on the good site (e.g., 9° compared with 3.5° at 150-mm diameter). However, most of this difference was due to the greater age required to reach the same diameter on the poor site (50 years compared with 20 years).

All simple correlations of tree age or tree radius with cumulative absolute spiral angle proved to be highly significant. However, when partial correlation coefficients were calculated, the effect of radius on spiral angle (with age held constant) was small and non-significant, whereas the effect of age on spiral (with radius held constant) remained significant. In this instance, therefore, age appeared to be the controlling feature, and growth rate merely an associated characteristic. Nevertheless, Kennedy and Elliott recommended that from a silvicultural point of view, trees should be grown as fast as possible so that they would contain an actual spiral angle as small as possible by the time they reached a merchantable diameter.

Elliott (1958) studied *Pseudotsuga menziesii* and *Tsuga heterophylla* growing in mixed stands in three areas with a wide range of site quality. At age 50 years *Pseudotsuga* had the same cumulative absolute spiral of 4.25° on the two better sites, but 7.25° on the poorest site. *Tsuga* was similar, though differences were less marked, 5.25° and 5.75° on the better sites and 6.5° on the poor one. Thus, in each species the poor sites with slowest growth produced the greatest cumulative absolute spiral angle, and, incidentally, also the highest actual average spiral. These results were confirmed by Wellwood and Smith (1962).

Elliott also found that the relative importance of age and growth rate also differed between sites. On the slower growing sites the change in spiral angle was highly correlated with tree age, but on the better sites stem radius exerted a stronger influence than age, although age was also strongly correlated. The difference is important, because it might be suggested that growth rate proved most effective where it was given most opportunity to function, i.e. on good sites that did not restrict tree growth.

Paul (1955) reported that spiral angle was reduced by accelerated growth in *Pinus ponderosa* that had been released, yet in the following year he also recorded that dominant trees of the same species had larger grain angles than codominant and intermediate trees at the same age (Paul 1956). This would seem to imply that accelerated growth rate has the opposite effect to vigour of growth as expressed throughout the life of a tree.

The results reported for *Pinus patula* by Paterson (1967) suggested that there was a weak negative relationship between growth rate and spiral angles, whether measured underneath the bark or as weighted mean values. However, mean angle at breast height compared with diameter "did not confirm this convincingly".

In his summary of studies in which slow growth was said to have favoured increased spirality, Sachsse (1965) referred to two studies, previously mentioned, in which poor site conditions had contributed to slow growth. The study of Raunecker (1957) attributed spirality to soil chemistry. That of Burger (1953) implicated poor site conditions with increased spirality in *Quercus robur, Carpinus betulinus, Aesculus hippocastanum*, and *Fagus sylvatica*. One must also include the numerous examples of trees at high altitude which develop severe spiral grain (Sect. 4.1.6). These certainly have lower growth rates, both in diameter and height, than their more favoured relatives in lowland forests.

Few studies have considered the effect on spirality of height growth alone, although the defoliation experiment of Harris (1969), already referred to, was said to have increased spiral angle in shortened branch internodes.

4.2.2 Fast Growth, Increased Spirality

Many workers have associated fast growth with wide spiral angles. For example, Braun (1854) remarked that open-grown trees were more inclined to spirality than trees grown in closed stands, and this view has been supported by others (Champion 1924, Cahn 1931, Jacobs 1935, Brazier 1967).

Jacobs (1935) believed that spiral grain in *Pinus radiata* was confined to the crown zone, and that open-grown trees which retained a deep live crown therefore developed spiral grain more than trees in closed stands, in which the dead level of the crown rises more rapidly. Even so, he also stated that spiral grain usually disappears before the stem has attained a diameter of 150 mm, so that its presence within crown-formed wood would seem to be somewhat tenuous.

In a rather unusual study, Turnbull (1942) used the strength ratio of wood tested in static bending to assess the presence of spiral grain in *Pinus pseudostrobus*. The ratio Modulus of Rupture/(Basic Specific Gravity)^x was investigated for straight-grained wood in which x was found to be 1.25 and the whole expression gave a value of 28000. This was followed by tests of 642 specimens with spiral grain included, but with other sources of cross-grain excluded. Any reduction of the strength ratio was therefore attributed to the effects of spirality. From these results, using even-aged stems growing on a uniform site, Turnbull concluded that spirality varies directly with rate of growth.

Rault and Marsh (1952) considered the effects of both height and diameter increment on spirality in *Pinus roxburghii*. A positive correlation was found between spiral angle and annual height increment ($r = 0.24$), but in view of the relationship between height and diameter growth they did not believe that the effect of height on spirality was necessarily a causal relationship. The correlation between spiral angle and diameter growth (expressed as mean annual increment in d.b.h.) was also positive ($r = 0.25$), as was the relationship between d.b.h. and spiral angle in trees of the same age. This latter relationship was further examined by means of regression coefficients. It was found that in stands of low mean spirality the relationship was weak or absent, but that it became stronger and more significant as mean spirality increased. They concluded "In other words the dominant trees tend to exhibit greater spirality than suppressed ones, and this tendency becomes more marked in stands of high mean spirality. Since the trees examined were growing in even-aged stands, this constitutes a strong indication that rate of growth is a factor influencing spirality irrespective of any supplementary effect of age".

By examining trees of the same seed source growing on different sites, Rault and Marsh came to the conclusion that trees which have been grown slowly may contain considerable resources of latent spirality, which may not be apparent unless age and rate of growth are taken into account. Such trees cannot be heavily thinned without increasing their spirality considerably, and they are quite unsuitable as sources of seed.

In a genetic survey of *Pinus radiata*, described in Section 6.4.2, Burdon (1981) found that spirality showed a positive within-population correlation associated with the width of the annual growth layer, but this was not strong. Finally, three further writers who found a significant positive relationship between diameter growth and spiral grain are Balodis (1960) for *Araucaria cunninghamii*, Brazier (1967) for *Picea sitchensis*, and Smith (1973) for three varieties of *Pinus caribaea* (vars. *bahamensis, caribaea,* and *hondurensis*). In fact Smith reported that spiral grain became worse with faster growth whether judged on the basis of individual trees, locations, or varieties, though the varietal differences were considered to be of little practical significance.

4.2.3 Spirality Independent of Growth Rate

The commercial importance of internodal length in *Araucaria cunninghamii* prompted Balodis and Sheard (1959) to see whether selection of trees with vigorous height growth had any effect on spiral grain. They concluded that there was no such relationship, and that selection of trees to maximise height growth would have no adverse effect in this respect.

Very extensive studies failed to indicate any relationship between rate of diameter growth and spirality (other than that attributable to age) for *Picea abies* (Krempl 1965b, 1970) or for *Larix kaempferi* (Mikami et al. 1972, Mikami 1973, Mikami and Nagasaka 1975). Webb (1967) also failed to discover any effect of growth rate on development of interlocked grain in *Liquidambar styraciflua*.

Nicholls (1971c) studying *Pinus pinaster*, and Cown and McConchie (1981) studying *P. radiata* were unable to detect any effect of thinning and fertiliser application on spiral grain, although both species had shown significant responses in radial growth increment. Clonal *P. radiata* trees were used by Nicholls and Fielding (1965) to study the effect of growth rate on a range of wood properties including spiral grain, but statistical analysis showed no significant difference between the slow- and fast-grown members. A slightly different approach was used by Nicholls and Brown (1973) who examined trees of *P. radiata* that had forked near ground level giving rise to stems with different growth rates, in effect "clonal pairs", with a common root system and an almost identical environment. Again, no effect of growth rate on spirality could be demonstrated.

4.2.4 Other Consequences of Growth Rate

Of other effects on grain direction attributed to growth rate, one of the more curious observations is that by Toker (1952) who reported that fast-grown trees of *Eucalyptus rostrata* tended to have interlocked grain, whereas slow-grown trees often had wavy grain.

Wellner and Lowery (1967) also suggested that growth rate influenced grain direction, in that trees grown in open stands (with large crowns and rapid growth rate) were reported as having LH spiral grain, whereas trees grown in very dense stands had RH spiral. However, Wellner and Lowery were reporting on pole-sized material, and it is possible that the fast-growing poles were younger and therefore still retained the initial LH spirality common to young conifers, whereas the slow-grown poles were old enough to have the reversed grain angle.

The consequences of there being some accepted "commercial diameter", such as the sizes acceptable for poles in the foregoing example, may be important for mean spirality. For example, Kennedy and Elliott (1957) recommended maximising growth rate of *Alnus rubra* because spiral angle increased with age. Therefore the quicker trees could be grown to commercial size, the smaller the mean spiral angle within them was likely to be. However, the situation is different for trees in which spiral angle reaches a maximum value in wood near the pith, and then tends to decrease with age. They will contain a high proportion of wood with spiral grain if grown rapidly to a given diameter. This is one consequence of growing trees such as *Pinus radiata* on short rotations (Nicholls 1973).

4.3 Balance of Controls

It is very difficult to isolate growth rate from other influences such as age, genetics, and environment. Elliott (1967) made a plea for standardised sampling procedures, and for application of the same strict types of analysis to be applied to spiral grain as have been applied to other wood properties, such as wood density and cell dimensions. Unfortunately, spiral grain seldom shows the same regularity as these other properties, yet there have been many studies, especially within the last 30 years or so, that go a long way towards meeting the most stringent requirements of sampling and analysis.

So, although the effects ascribed to growth rate in many of the examples quoted above may more properly be attributed to other factors, there remains a considerable body of evidence that growth rate can, by normal statistical criteria, affect spiral grain either positively, negatively, or not at all, according to the species and circumstances under test. The cynic might argue that these represent the one test in 20 that will meet the criterion $p < 0.05$ even in a random array, and that the other 19 have not been considered interesting enough to report. However, since all categories, including that of nil effect, are equally represented, it seems more rational to ask how such a diversity of responses might arise.

To maximise growth rate in a stem, requires a combination of optimal features, including supplies of carbohydrate and growth substances (which will depend on external conditions such as day length, temperature, water and nutrient supplies) and, at the same time, developing organs in other parts of the plant must be exerting minimal demands for their share of carbohydrate production. The timing and balance of these requirements will inevitably differ between species, and evidence has already been presented that individual species will respond differently to growth rate when any of these balances are changed, e.g., as summarised above:

1. *Pinus ponderosa* responds to a sudden acceleration in growth rate differently than when the same growth rate is maintained over an extended period.
2. In *Pseudotsuga menziesii* and *Tsuga heterophylla* spirality is more restricted by growth rate on good sites than on poor sites.
3. *Pinus roxburghii* responds positively to growth rate so that dominant trees are more spiralled than suppressed trees, and this tendency is more marked in stands of high spirality.

To understand what factors influence spiral grain and how they interact, is a task for many disciplines including plant anatomy, genetics, and physiology. These are issues that will be addressed in the following chapters, but it is appropriate here to indicate the types of problems they will be required to solve, and the necessity for looking at fundamentals, rather than relying on superficial observations.

4.4 Effects of Silviculture

Meanwhile the silviculturist has his forest to tend, and cannot wait for total understanding of the physiology and genetics of his crop. It is always sound practice to select seed from the best looking trees, thereby excluding those that can be recognised as containing spiral grain, but from this point onwards recommended cultural procedures are conflicting, to say the least.

Recommendations regarding control of growth rate by such means as initial spacing, thinning, or fertiliser applications will depend on the pattern of spirality (e.g., increasing or decreasing from the centre outwards) and the expected response for each species. Thus recommendations to restrict spiral grain by maximising growth in some species, but to hold growth rates down in others, though based on reliable observations, certainly provide no basis for generalisation.

Jacobs (1935) suggested that the way to reduce spirality in *Pinus radiata* was to lengthen the rotation, and to thin in order to obtain good diameter growth only after grain angles straighten out. In this way spiral grain could be restricted to a small central core which could be discarded. However, restricting spiral grain has never been a primary objective of *P. radiata* silviculture, and certainly does not figure in recent prescriptions.

Extreme response in growth rate, of the type associated with sudden release of suppressed trees, or where a soil mineral deficiency is rectified, have sometimes caused concern that wood properties of conifers might revert to those associated with wide rings in the region of the pith. The basis of this concern is that accelerated growth may temporarily reduce wood density and fibre (tracheid) length. However, in the writer's experience, there is no evidence that spiral grain is promoted. In extreme cases, top-dressing of forests deficient in phospate with superphosphate fertiliser, has suddenly changed spindly moribund *Pinus radiata* into vigorously growing trees. Within 2 years ring width has increased from 1 to 10 mm, without increase in spiral angle (unpublished data).

Gerischer and de Villiers (1963) reported that very heavy pruning, in which crown depth was reduced to one-third or less, reduced spiral grain in *Pinus radiata* for about 10 years. A significant correlation was found between spiral angle and growth rate in the pruned, but not in the unpruned trees.

If severely spiralled trees can be detected it is desirable that they should be removed in thinning. However, as Mayer-Wegelin (1956) pointed out, many species develop spiral grain late in life and are therefore not easily disposed of in this way. In theory it should be possible to detect badly spiralled trees at an early stage in species like *Pinus radiata* in which the initial LH spiral is of most concern, or in *P. densiflora* for which early spiral development is said to be a good indicator of mean angle at maturity (Nakagawa 1986). An instrument like that described by Pelle (1967), which can detect spiralled wood through thin bark, would lend itself to this use. However, to the writer's knowledge spiral grain has never been used as a criterion for thinning, even in seed stands or other high-value crops.

4.5 Advantages for Survival

Strength tests of models and actual stems with spiral structure led Thunell (1951) to suggest that trees with spiral grain in stem and branches are better equipped to withstands high winds and heavy loads of wet snow (Sect. 3.3.2). However, Krahl-Urban (1953 a, b) referring to *Fagus sylvatica*, and Venkataramanan (1967) referring to *Pinus roxburghii*, believed that spiral grain made these species less able to withstand heavy snow loads, although neither author presented any detailed evidence. Certainly, it is difficult to imagine that the severe spiral grain, which typifies many high altitude species at the upper limits of their height range, can be disadvantageous, otherwise this feature could hardly survive in so strenuous and demanding an environment.

Vité (1958, 1967) suggested another advantage for trees with spiral grain. They have the capacity to distribute the transpiration stream around the stem, so that every branch has direct access to almost every root by way of one or more of the growth layers involved in water conduction. This is not to deny the possibility of lateral water movement if need arises, but distribution of water by the easiest route, in the direction of the grain, is seen as an advantage in making maximum use of the root system. It would be particularly valuable for trees that face greater transpiration stress on one side of the crown than on the other, or which may have water procurement more limited to one side of the root system. It could also be of relevance in maintaining a balanced water supply, and a well-distributed demand on all roots, as requirements from different parts of the crown change diurnally, or with changing wind direction, and so on.

4.6 Conclusions

1. The ways in which the environment is supposed to influence the development of spiral grain have elicited some of the most fanciful and extravagant hypotheses concerning tree growth, but rather less factual information.
2. There is no support for the belief that regular patterns of spirality can be initiated by, or attributed primarily to, any environmental factor.
3. It seems probable that the propensity for spiral grain is determined by heredity, although its expression may be dependent, at least in part, on the environment.
4. The general unpredictability of environmental effects on spiral grain points to the need for better understanding of its physiological and genetic origins, as well as the anatomical processes that lead to its development.
5. Varied responses even within a single species make it difficult to offer foresters precise prescriptions for minimising spiral grain. Nevertheless, foresters should be aware of the needs for utilisation so that they can remove trees that fail to meet these requirements, and they should understand the normal patterns of spiral development with a view to maximising growth when spiral angles are smallest if this is possible.
6. It has been suggested that spiral grain has advantages for tree survival under some circumstances. However, in normal production forests straight-grained trees concede nothing in competitive ability, and no disadvantages can be seen in seeking to minimise spiral grain.

Chapter 5

Anatomy of Changing Grain Angles

5.1 The Role of Cell Structure

Studies of the structure and function of living cells and of the activity of protoplasm within them have led many workers to the conclusion that spirality is a fundamental tendency within the entire plant kingdom. Schmucker (1924), for example, reviewed right- and left-handed tendencies in plants, from bacteria and protozoa to pteridophytes and spermatophytes. In this respect he saw the development of spiral grain that had been described by Braun (1854) and Hartig (1895) as an example of the same natural trend.

This broad association between spiral grain and other spiral phenomena in nature has been inferred from such potentially related features as protoplasmic streaming (Champion 1924, Sinnott 1960), the spiral growth of certain unicellular plant organs (Preston 1949), and microfibril orientation in cell walls (Champion 1924, Misra 1939, 1943, Preston 1949, Sinnott 1960, Foulger 1966). Others have sought to invoke wide-ranging phenomena to elevate the spiral to some kind of driving force in nature. Seifriz (1933a, b) pursued this concept with such enthusiasm that he included not only molecular spirals, and spiral plant structures such as cotton fibres, tendrils, and vines, but also shells of molluscs, and mammalian organs, culminating triumphantly with human viscera and the cardiac loop. Even the tendency of people when lost to wander in circles, and of dogs to turn around before going to sleep were included in this catalogue of phenomena related to the "fundamental heritable protoplasmic quality".

Such generalised abstractions apart, it is necessary to understand the structural changes that give rise to spiral grain and also their underlying mechanisms. Any explanation must be able to account for features such as reversal of spiral direction and considerable developmental diversity at both intra- and interspecific levels.

Van Iterson (1943) attempted to develop a model that would explain spiral development, and he set a number of requirements for any such system. These can be summarised as follows:

1. The spiral structure must be developed in the absence of torque forces on the stem as a whole.
2. The system must be able to develop both RH and LH spirals.
3. The mechanism involved must relate to forces that are active in growing or moving organisms.

These are valid but demanding requirements. They will serve as a basis for this chapter, which, although concerned only with anatomical aspects, will leave the way open to consider in later chapters the other driving forces in living organisms, i.e. their genetic constitution and physiological requirements.

In particular, the structure and function of cambial cells will be of primary concern, because it has become apparent over the years that grain angles in wood reflect the orientation of cambial initials. It is in the cambium that the major changes take place. This is not, as Braun (1854) suggested, a passive process arising, as it were geometrically, from re-arrangement of initials to accommodate their elongation with increasing age. Changes in orientation of cambial cells can be related to a whole range of active cellular processes. Before considering these, however, it is necessary to look first at the vascular cambium, its structure and function, and the methods available to study changes within it.

5.2 Cambium

The lateral meristem known as the vascular cambium is not uniform, but consists of cells which perpetuate the meristem (cambial initials), and their undifferentiated derivatives which are phloem mother cells to the outside and xylem mother

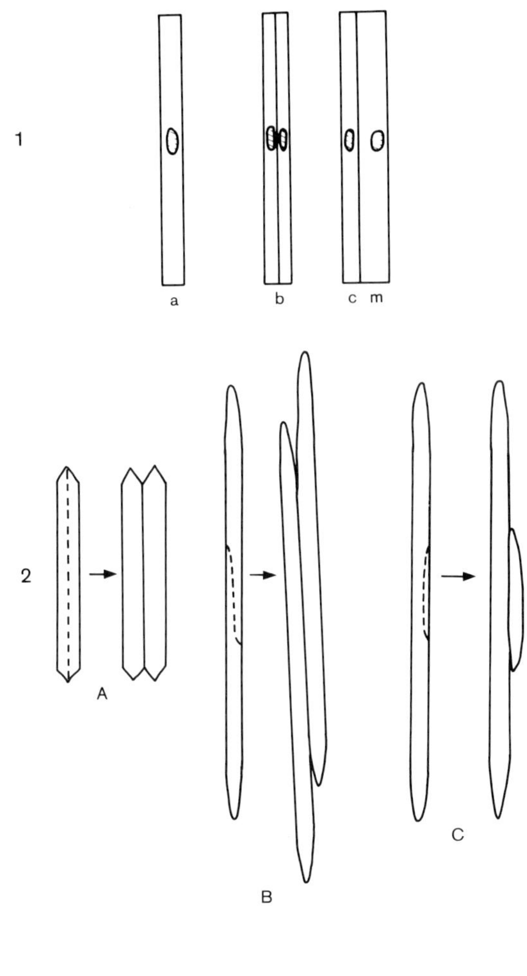

Fig. 5.1. Cell division in fusiform cambial initials. **1** Diagram of periclinal division in a fusiform initial (radial view). *a* Cambial initial; *b* tangential wall divides the cell equally; *c* one daughter cell continues as cambial initial, *m*. The other daughter cell enlarges as xylem or phloem mother cell. **2** Diagram of three types of anticlinal division and production of daughter cells (tangential view). *A* Radial longitudinal; *B* pseudotransverse; *C* lateral division

cells to the inside. Both xylem and phloem mother cells may undergo further periclinal divisions, and sometimes anticlinal divisions as well, but anticlinal divisions creating new cambial cells are restricted to cambial initials (Fig. 5.1). The whole region of undifferentiated cells is termed the cambial zone.

5.2.1 Methods of Studying the Cambium

Since the advent of electron microscopy cambial cytology has become much more accessible to study, and adaptations of the fixation, embedding, and sectioning procedures used in electron microscopy have been widely used to enable cambial cells to be studied directly using light microscopy (e.g., Barnett and Harris 1975). However, most of the information on orientational changes occurring in the cambium has been obtained from the record of these changes in the secondary xylem. Some workers have used the secondary phloem, but results are inevitably restricted with respect to the length of the record available, and may be misleading in some circumstances (Philipson et al. 1971).

From time to time various workers have examined the sequence from cambial initials to mature xylem elements to ensure that the xylem cells used in their studies provide a true indication of events in the cambium [e.g., Bannan (1950), using *Chamaecyparis thyoides* and *C. nootkatensis*; Hejnowicz (1961), on *Larix decidua*; Krawczyszyn (1971), on *Platanus acerifolia,* and Włoch and Zagórska-Marek (1982), on *Tilia cordata*].

Events in the cambium of gymnosperms (*Ephedra* and *Gnetum* excluded) can be safely inferred from changes in tracheid numbers and orientation as seen in serial tangential sections of secondary xylem, because tracheid elongation during differentiation is relatively small (5–10%). Serial sections with a thickness approximating the width of a tracheid (20–30 µm) can be compared, using some readily identified feature such as a group of rays to locate and observe division and re-alignment along radial files of tracheids. The uniform structure of gymnosperm wood allows the study of the consequences of virtually every cambial division in this way.

On the other hand, fibre elongation in dicotyledonous woods is often much greater than elongation of tracheids in gymnosperms and, in addition, the more complex structure of hardwoods precludes complete continuity in observations of cambial changes. However, several methods have been devised to permit observations at intervals short enough to allow useful comparisons. Where wood structure includes tangential parenchyma bands, these may provide tangential sections, uninterrupted by vessels, which can be used to observe periodic structural changes [e.g., Hejnowicz and Zagórska-Marek (1974), in studying *Entandrophragma*]. Hejnowicz and Romberger (1973) used the last-formed region of the annual growth layer, which they called "terminal late xylem" in their studies of *Acer, Betula,* and *Fraxinus.* In other species, for reasons that will be explained later (Sect. 5.3), changes may be followed through orientation and relative positions (division and recombination) of wood rays, as in studies of *Platanus* by Krawczyszyn (1972, 1973, 1977). Finally, Webb (1967) used the orientation of vessel perforation plates as viewed on the end grain of *Liquidambar* to assess the direction of cell division in relation to changing grain angles.

5.2.2 Nature of the Cambium

There is still some division of opinion as to whether there is a single permanent initial cell in each radial file of the cambial zone, or whether there is a multiseriate cambium in which all cells are equal. Currently, the concept of a single initial is most widely accepted (Philipson et al. 1971). However, there is insufficient evidence that a single initiating layer always exists in dicotyledons, and so, to some extent, the issue must remain open.

If the initiating layer were more than one cell wide in conifers, it would be difficult to explain the widely observed duplication of changes on either side of the cambial zone, i.e., doubling or loss of radial files, and replacement of fusiform elements by rays (Bannan 1968a). However, even when there is a single cambial fusiform initial in each radial file, the initials may not constitute a continuous layer at all times during active growth. Instead, they may take the form of an irregular network, from which parts of cells may project inwards or outwards from their adjacent initials (Włoch 1981). A continuous layer is, however, re-established as the cambium approaches dormancy.

Although re-orientation of a multiseriate cambium (if this exists) would appear to be more complex than re-orientation of a single initial, the difference is only one of degree. The central cell or cells in a multiseriate cambium must give rise to other initials that can inherit spatial adjustments in the same way as a xylem or phloem mother cell, the main difference being the presumed ability of a cambial cell towards the outside of a sequence to "correct" or add to any deviation. In this respect it could be argued that whereas "accidental" deviations in the centre could be corrected by exterior cells, deviations initiated by persistent and directed forces within the cambium might be responded to cumulatively (and therefore more rapidly) in a multiseriate cambium than in a single initiating layer. Probably the difference is unimportant: there is no reason to believe that the potential for cambial re-orientation would be significantly different in either case.

5.2.3 Cell Divison

Cambial cells may divide in a variety of ways. For reasons that will become apparent later (Sect. 5.2.6), most interest attaches to division of fusiform initials because they play a more active role in cambial re-orientation than the ray initials.

5.2.3.1 Anticlinal Division

Anticlinal division in species with non-storeyed structure is characteristically pseudotransverse (Fig. 5.1, 2B). That is, the dividing wall is usually laid down near the centre of the cell, but is oblique to a greater or lesser degree. In conifers the length of the dividing wall (or degree of obliquity) is directly correlated with cell length (Bannan 1964a, 1965a). However, the length and pitch of the wall can be very variable, ranging from short and nearly transverse (Kohl 1933) to very oblique, extending up to half of the cell length in conifers (Bannan 1957b) and even longer in some dicotyledons (Cumbie 1967).

Occasionally, anticlinal division does not extend from one side of the cell to the other, but cuts off a small cell from one side of a fusiform initial (Fig. 5.1, 2C). Cells formed in this way in gymnosperms seldom persist as fusiform initials, and if they survive are more likely to become ray initials. Actually, this type of division is rare in gymnosperm cambia (Bannan 1950, 1951a, 1957b, 1968a, Hejnowicz 1961), amounting to only about 1% of the total anticlinal divisions (Bannan 1968a). However, Hejnowicz (1961) noted a high frequency of lateral divisions in a root of *Larix decidua*. A higher frequency of lateral divisions has also been recorded in various dicotyledons, and in non-storeyed species with short initials the greater proportion of all anticlinal divisions may be lateral and/or truly radial/longitudinal (Cumbie 1967).

Most attention has been directed towards the cumulative effects of unidirectional pseudotransverse anticlinal division, followed by intrusive growth of the daughter cells, as a cause of cambial re-orientation since this possibility was first mooted by Hartig (1895). The consequences of pseudotransverse division are certainly easy to follow in the xylem, and daughter cells can be seen to take up an alignment at some angle to the mother cell. However, even the most rapid changes in grain angle require shifts of only a fraction of a degree in the axial directions of successive cambial initials (Preston 1949), so that the possibility of more subtle shifts in alignment has drawn more attention in recent years to changes following periclinal division (Harris 1973, Savidge and Farrar 1984).

Indeed, for storeyed cambium, anticlinal division and cell elongation would seem to provide few opportunities for re-orientation of fusiform initials because storeyed initials usually divide anticlinally in the radial longitudinal direction (Fig. 5.1, 2A). However, this may depend on how strictly the storeyed structure is maintained. Zagórska-Marek (1984a) reported pseudotransverse divisions, intrusive elongation, and a considerable frequency of cell failure in the relatively flexible storeyed structure of *Tilia* cambium.

5.2.3.2 Periclinal Division

After periclinal division (Fig. 5.1, 1B), one of the daughter cells retains its function as an initial, expanding radially before dividing again. If there is a single initiating layer the other daughter cell then becomes a xylem or phloem mother cell. It too expands radially, and may or may not undergo further division before differentiating into a mature vascular element. Further mitosis is, in any case, usually delayed until one or more undifferentiated derivatives lie between it and the initial cell (Bannan 1950, 1955).

Although periclinal divisions usually produce daughter cells of equal dimensions, this is not always so. Division may be partial, not extending from one end of the cell to the other, so that the cell retaining the function of initial becomes shorter (Dodd 1948, Bannan 1953). This has been termed imperfect periclinal division by Savidge and Farrar (1984). Sometimes the cell plate may appear to be warped, even to the extent that it appears periclinal in one part of the cell and anticlinal in another (Włoch 1981), in which case an initial may change its shape and, to some extent, its orientation.

Imperfect periclinal division, and uneven expansion between periclinal divisions are two additional processes whereby fusiform initials in non-storeyed cambium may have the potential to change their orientation. In strictly storeyed cambium they would appear to be the most likely, perhaps the only, means by which fusiform initials can change direction (Hejnowicz and Zagórska-Marek 1974).

5.2.4 Cell Survival

Priestley (1930) calculated that no more than four anticlinal divisions per initial would be needed to accommodate circumferential expansion during 60 years' growth of a pine stem. However, when the first actual observations of the frequency of anticlinal division were made in 1950, by Whalley (in *Thuja*) and Bannan (in *Chamaecyparis*), they discovered that the rate of division was far in excess of that required for circumferential growth, and that the majority of fusiform initials formed in this way failed to survive (Fig. 5.2). Bannan regarded fusiform initials as being competitive individuals in an overcrowded environment. These early observations led to the classic series of studies on the cambia of various gymnosperms by Bannan (1950–1968), which undoubtedly inspired much wider interest in the cambium elsewhere, including the work on spiral grain initiated by Hejnowicz (1964).

The rate of anticlinal division tends to be higher in young stems than in old ones (Bannan 1950, 1960b, Cumbie 1967). Three or four times as many divisions per growing season take place in initials of young trees as in older trees. The rate of survival is also greater in young stems, so that on average 4.5 divisions will add one fusiform initial to the cambium in a young stem, whereas it may require more than 40 divisions in an older but still vigorous one.

Bannan (1960b) also found that numbers of anticlinal divisions per growing season are affected by growth rate. In fast-growing stems there may be three or four anticlinal divisions per growing season, compared with one division over two seasons in slow-growing stems of the same age. When frequency of division is expressed in terms of the thickness of the wood (number of cells) produced, however, it tends to be more constant. For example, Bannan (1960a) found that cambial initials in *Thuja* stems more than 150 mm in diameter underwent an anticlinal division two or three times per centimetre of xylem increment regardless of tree size or rate of growth. Exceptions may occur in very young trees (Bannan 1960a), in very narrow growth layers (Bannan 1950, 1960a, 1962, 1963a, 1965a, 1966a), and in concavities of fluted stems (Bannan 1957a), in all of which the rate of anticlinal division tends to be relatively high.

The rate of loss of fusiform initials from the cambium is determined by the additional cells that can be accommodated as a consequence of circumferential growth, compared with cell production in terms of anticlinal division and elonga-

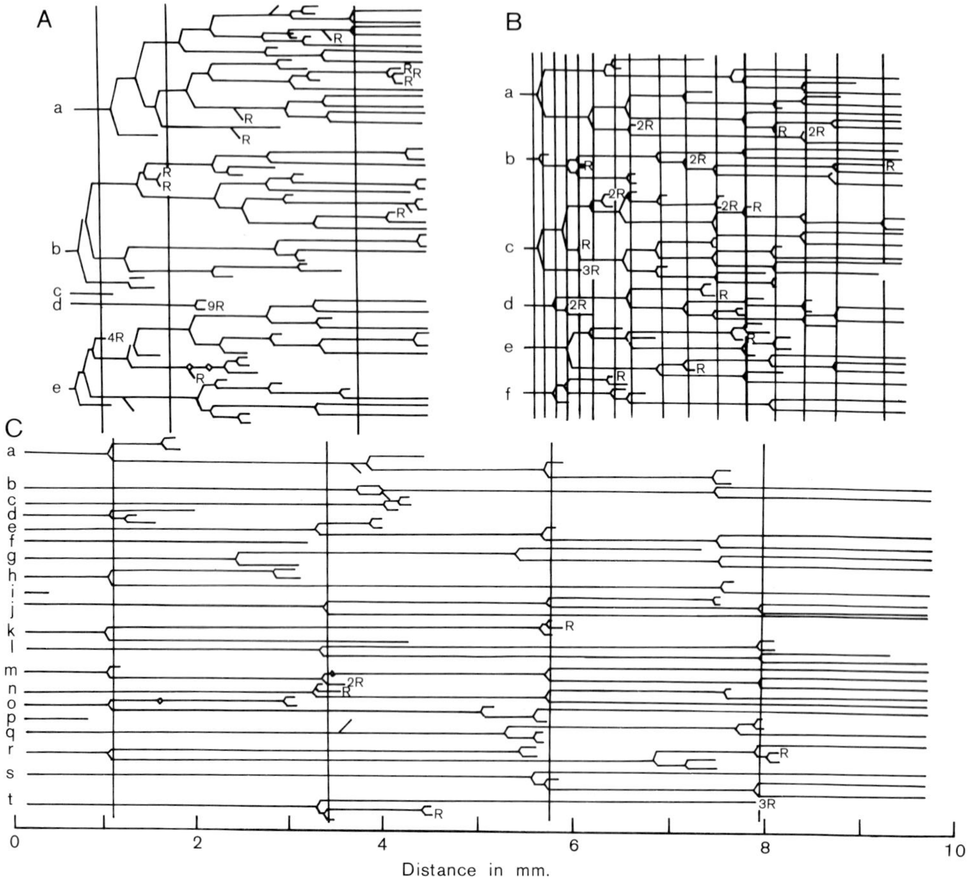

Fig. 5.2 A–C. Diagram of the succession of tracheid rows in selected lineal series, showing the relative time and frequency of anticlinal divisions. *Equal forkings* indicate centred pseudotransverse division; *unequal forkings*, eccentric pseudotransverse division; and *side branches*, lateral division. The *termination of a line* indicates loss of a fusiform initial, and the letter *R* indicates initiation of the number of ray cells given by the appropriate digit. **A** Young vigorous stem of *Chamaecyparis lawsoniana*. **B** Young slow-growing stem of *C. thyoides*. **C** Old, but vigorous stem. 22-cm diameter, of *C. thyoides*. (Bannan 1950)

tion of initials (Bannan 1962, 1963 a, 1964 a, b, Hejnowicz and Brański 1966, Hejnowicz 1967 a, b). Hejnowicz (1967 a) developed formulae from which it is possible to calculate the failure rate of new initials and the rate of intrusive elongation of surviving initials, for any given cambial radius, if mean cell length and frequency of anticlinal division are known.

The processes of decline in fusiform initials that fail to retain that function were described by Bannan (1950, 1951 a, 1953) and Whalley (1950). A declining initial becomes smaller in all dimensions. The rate of periclinal division is often retarded, and imperfect periclinal divisions become more frequent. Short cells produced in this way may persist in the cambium as ray initials, or periclinal division may eventually cease altogether, so that the initial loses that function and eventually matures as a xylem or phloem element. The radial file that in initiated is then obviously terminated.

Sometimes a fusiform initial may divide transversely, thus giving rise to one or several ray initials. Bannan (1953) found that in some conifers the relatively short initials of small stems or branches tended to subdivide only to the first or second order, whereas larger initials in mature stems usually subdivided to the third order.

Survival of fusiform initials may be under polar influence. Bannan (1968b) studied 28000 pseudotransverse divisions in 20 species of conifers, and found that in some areas of cambium the upper of two sister cells was more likely to survive, while in other areas the reverse was true. There was therefore considerable variation within a single tree, but there was also an overall tendency within a species for either lower or upper daughter cells to have a better chance of survival. When the tendency is for the upper cell to survive, cell elongation is predominantly acropetal, whereas basipetal elongation predominates when the lower cell survives.

Long fusiform initials have a better chance of survival than short ones (Bannan 1951b, 1957c, Bannan and Bayly 1956), and therefore the products of lateral division seldom survive as fusiform initials (Bannan 1957b). This may be because short initials have fewer contacts with ray initials (Bannan 1951b, 1957c, 1965b, Bannan and Bayly 1956). On average, surviving initials have contact with 30% more ray initials than those that fail.

5.2.5 Cell Re-arrangement and Elongation

In the dynamic and highly competitive environment of the cambium some cells fail to hold their ground against their more competitive neighbours and may eventually cease to exist. New wall contacts must then be made by fusiform initials on either side of the initial that has been lost. The files of tracheids derived from these convergent initials also have to develop matching pit pairs. That these developments are achieved regularly and very rapidly is an indication of the essential plasticity of the initiating layer and the adaptability of developing elements.

Cell elongation, which is achieved almost exclusively by intrusive growth, is another process that requires great adaptability by cambial initials and maturing vascular elements. After anticlinal division, the daughter cells in non-storeyed cambium elongate, often quite rapidly, so that those cells that survive as fusiform initials may reach or exceed the length of the parent cell. Details of the process of elongation have been described for *Chamaecyparis* (Bannan and Whalley 1950), and for *Thuja* (Bannan 1956). The process is, in fact, very variable, and often appears to be affected by surrounding cells as much as by any inherent capacity of the cell to expand. Sometimes intrusive elongation is able to proceed unhindered, in which case growth usually conforms to the normal pattern of a rapid initial phase followed by a gradual decrease in growth rate. However, the process is often subject to various hindrances.

A considerable delay in cell elongation may arise if the tip of an elongating fusiform initial comes up against a ray cell. During the delay in growth the cell tip often becomes flattened or forked, before the growing tip slides around (or through) the ray and continues to grow. Similar cell tip distortions arise when one fusiform cell meets another, and this applies equally to storeyed cambia (Włoch and Zagórska-Marek 1982).

Elongation of daughter cells after pseudotransverse division is at first more rapid in the overlapping tips (Bannan and Whalley 1950), but growth in the distal tips soon catches up so that their combined elongation equals that at the overlap, irrespective of any tendency for greater elongation in one direction. When elongation is slowing down in the final phase of growth, Bannan (1956) observed that cell tips in *Thuja* sometimes "recede", or fail to hold the space they have gained. If the tip at the other end then continues to elongate, the overall effect is one of vertical displacement or movement.

The manner of elongation of cambial initials was for many years the source of much discussion, mainly because of the apparent impossibility of having pits on adjacent walls retain their function if they slide past one another. However, it has been shown that pits are not necessarily dependent on cell wall organisation in the cambium, but can develop later as the cells differentiate (Barnett and Harris 1975).

Bannan and Whalley (1950) showed that growth of each fusiform initial in *Chamaecyparis* proceeded independently of its neighbours. Growth in length appeared to be limited to the ends of the initials, but it was impossible precisely to define the limits of the region involved. In radial sections of the secondary xylem, bordered pits on radial walls of tracheids were shown to be in approximate alignment in the centres of cells in a radial file but were distributed more irregularly towards the ends. It was thought that this irregularity probably reflected changing cell relationships due to elongation of the initials, although growth during differentiation may also have been responsible. Later workers (e.g., Hejnowicz and Romberger 1973) have, for a variety of reasons, also concluded that cell elongation is confined to the tips of fusiform initials.

Wenham and Cusick (1975) suggested that the tip of an expanding fibre secretes an enzyme which weakens the middle lamella of cells adjacent to it. The corners of these cells then round off from each other during cell expansion, creating a space into which the elongating tip can grow. Similar parting of the middle lamella ahead of growing cell tips had been observed by Neeff (1914). Hejnowicz (1980) demonstrated that there is tension across the radial middle lamellae in active cambia of both conifers and broad-leaved trees. He saw this tension as being important for intrusive growth of cell ends, and also for an increase in diameter of differentiating vessel members.

The situation in the xylem of trees with storeyed structure obviously differs from that in trees with non-storeyed structure. Cell elongation in storeyed cambia must play a very restricted role in cell re-orientation, yet many species with storeyed structure are notable for strongly interlocked grain (e.g., *Entandrophragma* spp.). Changes of grain inclination in storeyed cambia have been described for *Entandrophragma* and *Tilia* by Hejnowicz and Zagórska-Marek (1974), Zagórska-Marek (1975), and Włoch and Zagórska-Marek (1982). In suc-

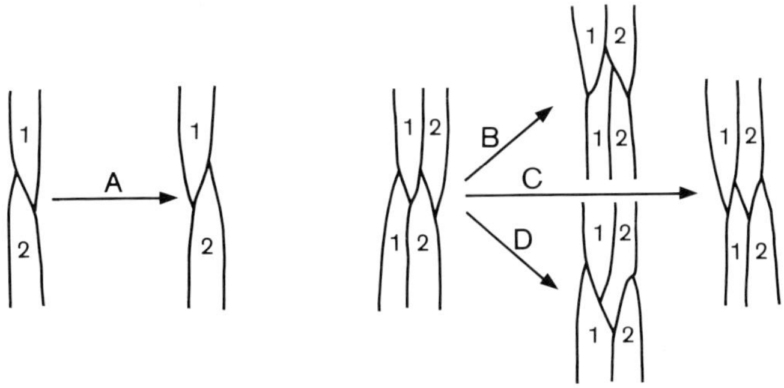

Fig. 5.3 A–D. Diagram of changes in the position of cell ends in storeyed cambium of *Entandrophragma*. **A** change of overlap by lower cell from *left to right*. **B–D** Progressive movements which may occur when two lower cells, which are changing from left to right, overlap. (Hejnowicz and Zagórska-Marek 1974)

cessive periclinal divisions the pointed ends of fusiform initials in a story change position relative to those of the storey above or below. This causes a sort of creeping movement by cell ends of one storey past those of an adjoining storey (Fig. 5.3). As one cell end moves past another it sometimes straddles it, becoming temporarily forked, before moving on to the other side. The opposite ends of cells move in opposite directions, so that each cell rotates about its centre as the angle between the fusiform initials and the stem axis changes.

5.2.6 Plasticity of the Cambium

Sinnott (1960) made the statement that "the cambial region is obviously a much more active and plastic one than early workers regarded it". It is, perhaps, timely at this point to review cambial plasticity, because it makes changes in shape and orientation of cambial cells possible to a degree that often seems incredible to plant anatomists more accustomed to the structural rigidity that derives from strongly cellulosic and lignified plant cell walls.

We have seen that the processes which may be involved in changing orientation along a sequence of cambial initials are:

1. Pseudotransverse division. According to Hejnowicz (1967 a, b) the change in grain angle is nearly proportional to the square of the frequency of pseudotransverse divisions when these are all in one direction.
2. Elongation by intrusive growth. Bannan (1966 a) and Hejnowicz (1973 a) found that this is often a strongly directed process.
3. Incomplete periclinal division. Savidge and Farrar (1984) believed that this could be the primary event leading to re-orientation of fusiform initials in some circumstances.
4. Differential growth after periclinal division (Harris 1973) and/or incomplete periclinal division (Savidge and Farrar 1984).

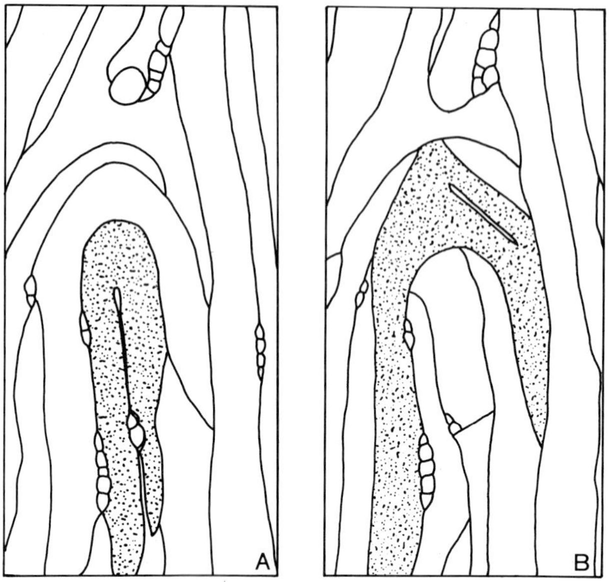

Fig. 5.4 A, B. Distorted tracheids in a wood tumour of *Picea excelsa*. **A** Tracheid (*stippled*) bent back on itself. **B** Tracheid with a self-enclosed cavity. Note also hooked and branched tracheids. (Włoch 1976)

There are circumstances in which these processes seem barely adequate to describe the course of events in the cambium. For example, the rotation of fusiform initials in storeyed cambium of *Entandrophragma* (Hejnowicz and Zagórska-Marek 1974) takes place without cell elongation, other than that necessary to meet the geometry of the situation. Initials derived from successive periclinal divisions develop a small angle with their predecessors, and these angles build up cumulatively to produce strongly interlocked grain. Incomplete periclinal division followed by differential growth that restores the shape of the initial may be the source of these changes, but Hejnowicz and Zagórska-Marek simply described it as "creeping" of cell ends in one storey past those of another.

Questions must also arise concerning abnormal cambial changes such as those following stem decapitation (Neeff 1914), formation of wood tumours (Włoch 1976), or spiral girdling (Harris 1973). Under these circumstances fusiform initials may develop hooked, branched, and other distorted shapes, twining around and thrusting between other fusiform and ray initials (Fig. 5.4). It is conceivable that these distortions could arise by the processes that have been described if taken to their extremes, but they certainly arise very rapidly, on the fringes of wound tissue, and in response to changes in stem polarity generally.

In both respects – slow changes of orientation without significant cell elongation, and the rapid changes during abnormal growth – the end results convey the impression of an almost amoeboid movement by fusiform initials. It is as if the ultimate shape adopted by the gelatinous envelope that is the primary wall were dominated by the demands and stresses of the cytoplasm within it.

It is generally agreed (Jones 1963, Hejnowicz 1972, Harris 1973) that ray initials play an entirely passive role (or are, at least, not mechanically involved) in changing grain angles. Fusiform initials are the active agents. Their activities cause groupings of ray initials to rotate about their axes, to be split, or to recombine, but through all this manipulation wood rays normally retain their positions relative to the stem axis to a remarkable degree.

There are, therefore, many types of events that can contribute to or reflect the changing orientation of cambial initials. However, when secondary xylem is used to study cambial changes, some of the processes of change are found to leave behind a clearer record than others.

5.3 Morphogenetic Events

The expression morphogenetic event (also sometimes morphogenic event) was first used by Hejnowicz (1972) to describe four readily recognisable changes available in the record provided by serial tangential sections of xylem. These are illustrated in Fig. 5.5. The first involves a change in the relative position of two fusiform initials, as indicated by the direction of overlap between their tips. Bannan (1966a) first commented on the directional nature of this type of event. The second records the direction of pseudotransverse division, and the third and fourth result from the "movement" of the ends of fusiform initials relative to ray initials. If the end of a fusiform initial grows through a grouping of ray initials the wood ray is split into two, and if the tip of a single fusiform initial that separates two groupings of ray initials is withdrawn, the two rays become united (Hejnowicz 1968). All of these events can be interpreted as favouring re-orientation of the elements involved either towards the right (Z event), or towards the left (S event).

Two difficulties arise in describing these events. It is convenient to refer to "movement" of cambial initials, but it must be realised that the expression commonly refers to changes in the relative positions of individual cells along a radial succession of initials. It does not necessarily imply some sort of jostling or wriggling by a maturing initial between its neighbours.

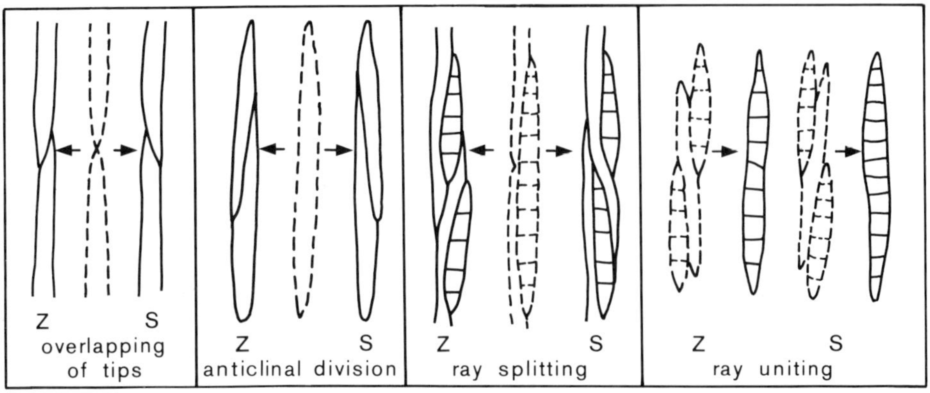

Fig. 5.5. Four types of morphogenetic events in cambial cells giving rise to RH (*Z*) or LH (*S*) grain angles. (Hejnowicz 1973a)

The second difficulty lies in the concept of causation. For the remainder of this chapter we shall be considering how the organisation and frequency of morphogenetic events bring about the changes of grain angle that lead to spiral, wavy, and interlocked grain. In this sense there is an apparent unity in the events leading to, or "causing", these phenomena. Nevertheless one should not lose sight of the possibility, clearly recognised at the outset of the studies initiated by Hejnowicz (1964), that a necessarily limited range of anatomical responses may arise from quite different directional stimuli. However, these issues lie in the realm of physiology and will be the subject of Chapter 7.

5.4 Cambial Domains

The slope of the dividing wall in pseudotransverse anticlinal division tends to develop in the same direction in neighbouring cells over a considerable area of cambium. This relative uniformity may extend over several millimetres or up to several metres (Bannan 1963a, b, 1964a, b, 1966a, b, Hejnowicz 1961, 1964, Krawczyszyn and Romberger 1980). At first Hejnowicz (1964) coined the term "cambial domain" specifically to describe areas of cambium in which anticlinal divisions were in the same direction. However, it soon came to be recognised that the direction favoured by other morphogenetic events coincided with the direction of anticlinal division, so all were included.

A cambial domain therefore came to be recognised as any area within which the greater proportion of the most readily recognisable morphogenetic events lay in the same directon, or favoured the change of grain angle in one direction. The presence of cambial domains in hardwoods, just as in softwoods, was then demonstrated by Hejnowicz and Krawczyszyn (1969) in their studies of *Acer, Aesculus, Alnus, Betula, Carpinus*, and *Liriodendron*. The actual morphogenetic events used in a particular study naturally depend on those that can be recognised in the species under investigation. For example, in his studies of *Platanus* Krawczyszyn (1971, 1972, 1977) had to rely heavily to the phenomena of ray splitting and uniting in terminal xylem of the annual growth layers.

Both Bannan (1966a) and Hejnowicz (1964, 1968) realised that predominance and persistence of morphogenetic events in one direction would probably result in changing grain angle and could induce spiral grain. However, they discovered that domains are not static. The orientation of events at any point changes with time. Grain would remain straight (or deviate only slightly in either direction) if the frequency of morphogenetic events were low, and if the periods tending in one direction were short and equal to those tending towards the other direction. For slope of grain to change significantly, the total number of events has to be high, and numbers in one direction must exceed the total in the other direction for a significant period.

The role of cambial domains in changing grain angles first became evident when Hejnowicz (1971) examined the morphogenetic events leading to wavy grain in *Picea abies*. The region of wavy grain in the tree he examined was very limited, measuring little more than 7 cm longitudinally and radially, and 5 cm tangentially. Through serial tangential sections it was observed that the cambial domains con-

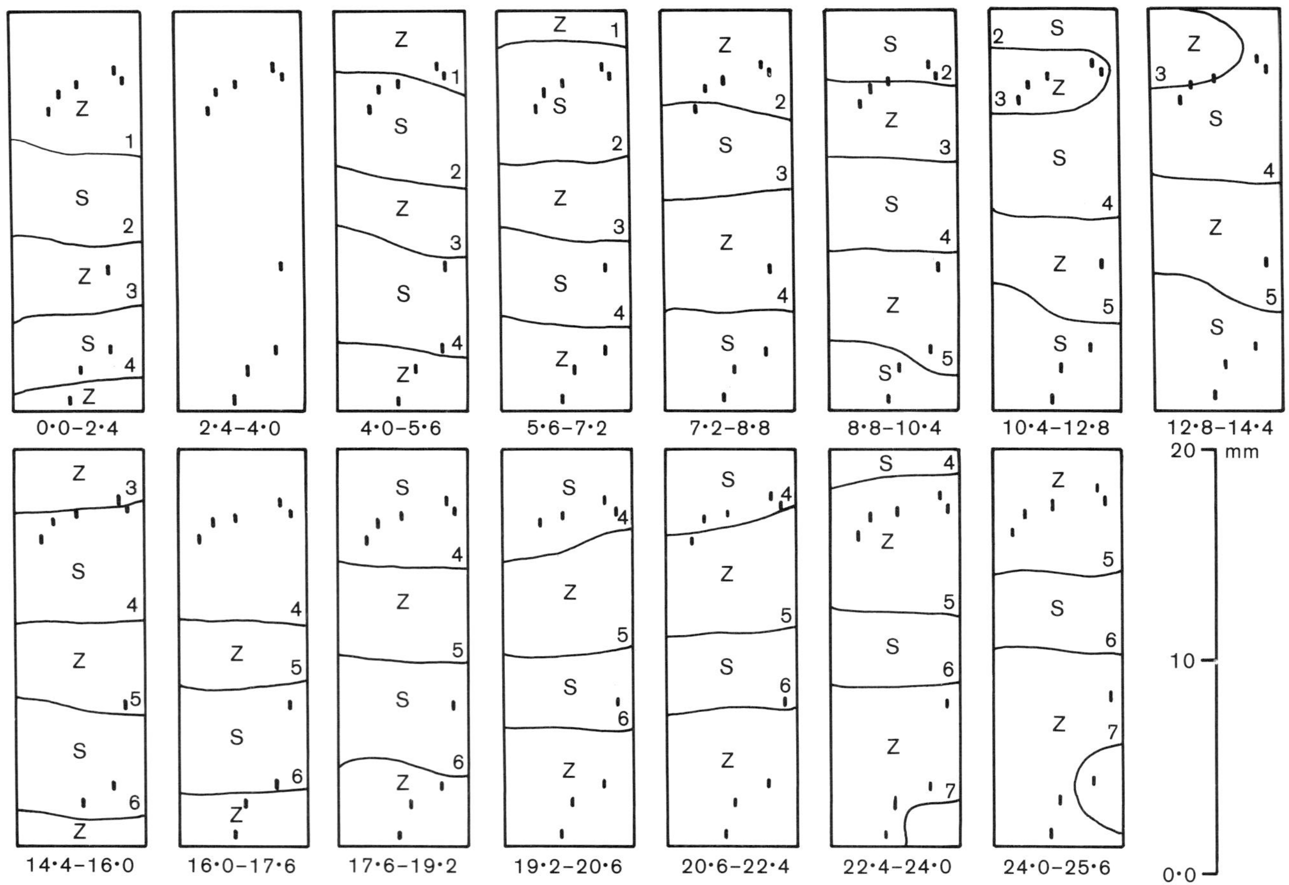

Fig. 5.6. Domain patterns during formation of wavy grain in *Picea excelsa*. Areas are lettered to indicate whether the majority of morphogenetic events within them tended towards the right (*Z*) or left (*S*), and show how the numbered domains moved upwards in successive sections. During the formation of wood from 2.4 to 4.0 mm no perceptible change in cell orientation occurred. Ray configurations used to locate sections are shown as *small vertical lines*. (Hejnowicz 1971)

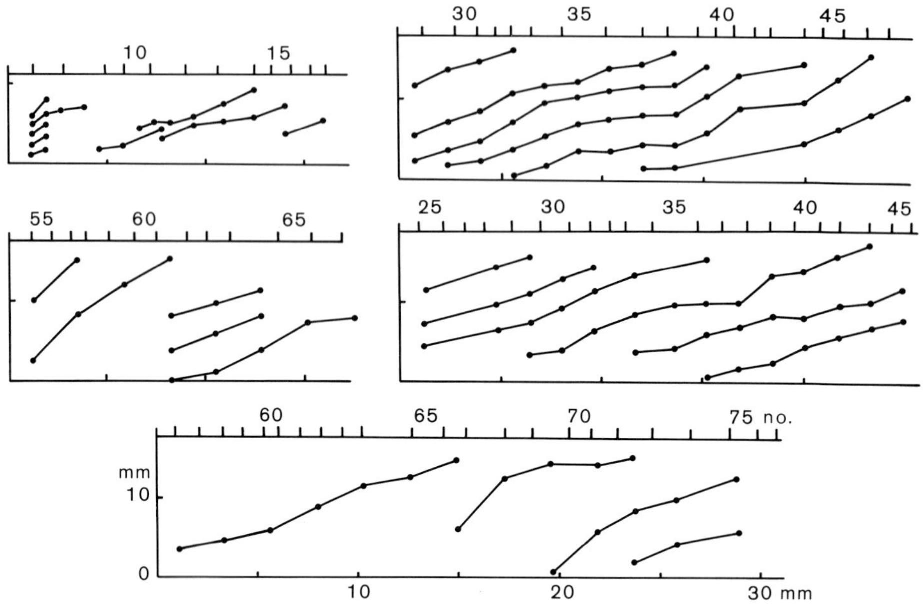

Fig. 5.7. Graphs showing movement of borders between domains in samples of wavy grain in *Picea abies*. The *lower horizontal axis* indicates radial distance, and the *upper* one ring number from the pith. The *vertical axis* gives longitudinal distance. (Hejnowicz 1971)

taining S events (tending towards development of LH grain angle) or Z events (tending towards RH grain angle) measured 5–10 mm longitudinally. These S and Z domains alternated down a tangential section through the zone of wavy grain, and when viewed on successive sections from the centre outwards the domains moved upwards, on average by 0.7 mm for each 1.0 mm of xylem radius (Fig. 5.6). The frequency of morphogenetic events was very high. The result was that the LH grain angle that had developed during the period when S events predominated, was first reduced in value and eventually became RH grain angle when that area became occupied by a Z domain. In other words, changes in grain angle across a radius reflected a succession of domains moving upwards. Naturally, boundaries between reversals of grain direction lagged behind areas occupied by domains, because of the time required to reduce a grain angle in one direction before reversal took place.

The rate of migration of the domains was also reflected in the angle between ripple marks as seen on a radial face (cf. Fig. 2.12) and a true horizontal (e.g., as represented by the wood rays). Hejnowicz (1971) illustrated this in another way, by graphs showing the movement of domain boundaries through a series of tangential sections (Fig. 5.7). These are consistent with domain movement of 0.7 mm per 1.0 mm of xylem added to the stem, or an angle to the horizontal of 35° for ripple marks on the radial face.

In 28 further examples of wavy grain from different trees of *Picea abies*, Hejnowicz observed that the wavy undulations ran upwards centrifugally at angles varying from 20° to 50°. In only one sample did the undulations run downward, suggesting that the domain pattern also moved downward during its formation. However, in the tree from which this sample was taken, other regions of wavy grain showed the normal upward inclination, so that Hejnowicz dismissed the downward movement as a "local phenomenon", or possibly an error in marking the sample.

In other studies domain movement has most commonly been found to be up the stem [e.g., Krawczyszyn (1972) for *Platanus*; Hejnowicz and Romberger (1973) for *Acer, Betula,* and *Fraxinus*; Zagórska-Marek and Hejnowicz (1980) for *Entandrophragma*]. However, domains that migrated basipetally were recorded in *Aesculus hippocastanum* by Pyszyński (1972), and quite rapid lateral movement of domains producing straight-grained wood was described by Hejnowicz (1972). The association of domain patterns with development of interlocked grain was first observed by Krawczyszyn (1972) in *Platanus*, and was later extended to a species with storeyed cambium (*Entandrophragman cylindricum*) by Hejnowicz and Zagórska-Marek (1974).

It is of interest that geometry requires fusiform initial length to increase with the angle of inclination when storeys are strictly maintained and their height is fixed. Krawczyszyn and Romberger (1979) confirmed this effect for *Entandrophragma cylindricum*, showing that the lengths of fusifom initials increased porportionally with the secant of the grain angle. This raised the question as to whether cyclical changes in the lengths of initials might be the cause of changing grain angles. However, while admitting that there was no a priori reason for disregarding this possibility, they thought it unlikely, because they did not find evidence of cyclical length changes in non-storeyed cambia producing interlocked grain (*Nyssa sylvatica* and *Platanus acerifolia*).

5.4.1 Domain Origins and Intensity

Domains can be detected early in stem formation. Krawczyszyn (1973) found evidence of domains within the first annual growth layer of *Platanus*. Transverse boundaries between domains coincided with the nodes, and the two lateral boundaries lay along the line of the major leaf trace and along its projection on to the opposite side of the stem (Fig. 5.8 A). A Z domain lay to the right of the upper leaf trace terminating an internode, and an S domain to the left. As further growth layers were formed, the size and shape of the domains changed, until by the fourth year no trace of the original nodal pattern could be detected.

Since domains can be detected in stems without marked development of spiral grain, it must be emphasised that the outcome is very dependent on the rate at which morphogenetic events occur within them. Insofar as anticlinal divisions and intrusive growth of fusiform initials are the fundamental features behind morphogenetic events, it follows that an area of cambium with high activity in this respect will contain fusiform initials that are somewhat shorter than normal. This is because cell elongation cannot immediately compensate for cell division (Hejnowicz 1972). Hejnowicz summarised these features for *Picea abies* (Table 5.1), in which areas of cambium actively producing wavy grain are compared with "neutral" areas at the margin of the same zone, and with an area producing nor-

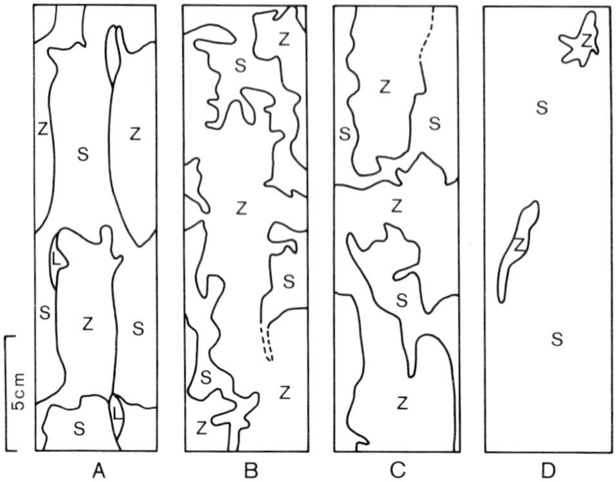

Fig. 5.8. Maps of domains (based primarily on ray splitting and uniting events) in *Platanus* during formation of the first four annual growth layers. *S* and *Z* domains are marked, and the positions of the major leaf traces (*L*) are also indicated on map **A**. Each map represents the entire cambial cylinder for each year at the same level. (Krawczyszyn 1973)

Table 5.1. Comparison of cambial events producing wavy and smooth xylem in the same tree of *Picea abies*. (Hejnowicz 1972)

Type of cambium	Mean frequency of anticlinal division	Mean rate of intrusive growth	Mean tracheid length
	(No. per cell per 1 mm of xylem)	(mm per cell per 1 mm of xylem)	(mm)
Active wavy	1.48	0.77	0.87
Neutral wavy	0.65	0.53	1.15
Normal smooth	0.45	0.40	2.49
	(No. per 1 mm cell length per 1 mm of xylem)	(mm per 1 mm cell length per 1 mm of xylem)	
Active wavy	1.70	0.88	
Neutral wavy	0.57	0.46	
Normal smooth	0.18	0.16	

mal straight grain some 20 cm removed from the wavy grain. Activities are compared in terms of number of morphogenetic events per initial cell per millimetre of xylem laid down, and also per millimetre of cell length per millimetre of xylem, in effect, per unit volume of xylem.

5.4.2 Domain Integrity

The orientational conformity within domains is such that they often contain less than 1% of "non-conforming" events (i.e., morphogenetic events tending to induce change of grain in the direction opposite to that induced by the majority). However, not all domains exhibit this degree of conformity at all times. Hejnowicz (1972) recorded a small coherent area containing non-conforming events within an otherwise uniform domain in *Picea excelsa*. It persisted for only 1 year, after which the major domain remained intact as if it had never existed. Such occurrences are of interest because they may imply that directional controls effective in the plane of the cambium can interact, temporarily at least, on a smaller scale than is normally apparent at the domain level.

In this respect it may be significant that morphogenetic events may not occur at a uniform rate even within a domain in which the orientation of events is uniform. Savidge and Farrar (1984) studied events in a diagonally girdled stem of *Pinus contorta* and in natural spiral development in *Picea glauca*. In both, homologously sloping pseudotransverse anticlinal divisions combined with imperfect periclinal divisions to bring about the observed changes in grain angle. In both examples it was also observed that the major changes were initiated by small groupings of adjoining fusiform initials, whereas the greater proportion of initials changed much more slowly. The small aggregates of cells responsible for the most rapid changes were termed "microdomains" by Savidge and Farrar. Even apparently uniform domains can therefore appear "patchy" when closely examined.

Włoch (1985) drew attention to long-term variations in the activity of cambial domains in hardwoods. In his own study of *Tilia cordata*, Włoch found that periods of activity sometimes alternated between domain types. In one sample

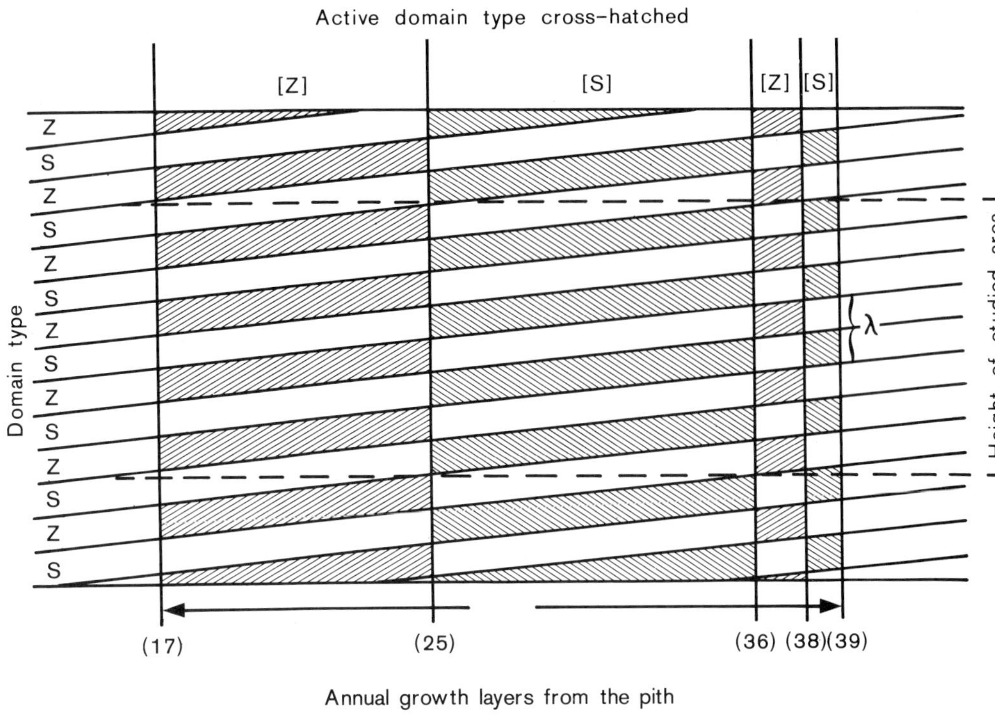

Fig. 5.9. Diagram of active and inactive domains in *Tilia* with interlocked grain. Wave period is 22 years, and during the time of one period the sequences of domain activity changed at 8-, 11-, 2-, and 1-year intervals. (Włoch 1985)

only Z domains were active during one period of 8 years, and this was followed by an even longer period (11 years) in which only S domains were active (Fig. 5.9). Nevertheless both types of "potential" domains (whether active or inactive) continued to migrate at a uniform rate. The result was interlocked grain with a wavelength much greater than the height of a pair of domains.

In reviewing previous work, Włoch (1985) pointed out that differences in frequency and distribution of events between Z and S domains could be detected in results presented by other workers (Hejnowicz 1968, Krawczyszyn 1971, Hejnowicz and Romberger 1973) and that periods of domain inactivity were clearly visible in descriptions of *Fraxinus* (Hejnowicz 1974) and *Aesculus* (Pyszyński 1977a), although little significance was ascribed to these features at the time.

Several workers have discovered similar zones of activity and inactivity when studying the creeping of fusiform cell tips in storeyed cambium (Hejnowicz and Zagórska-Marek 1974, Zagórska-Marek 1975, 1984a, Włoch and Zagórska-Marek 1982). This type of zoning, in which predominantly upper cell tips may be involved at one time, and predominantly lower tips at another, is not of the domain type with lateral or longitudinal shifts or activity. Rather, it is typified by somewhat random groupings of cells being active or inactive at different times. Włoch and Bilczewska (1987) refer to this condition as "fibrillation", presumably by analogy with cardiac fibrillation, though unfortunately confusing in a botanical context.

We have already seen that domain shape can change (Sect. 5.4.1). On the other hand, some domains are remarkably stable, producing continuous grain patterns over many years. Wavy grain, for example, often exhibits extreme regularity with

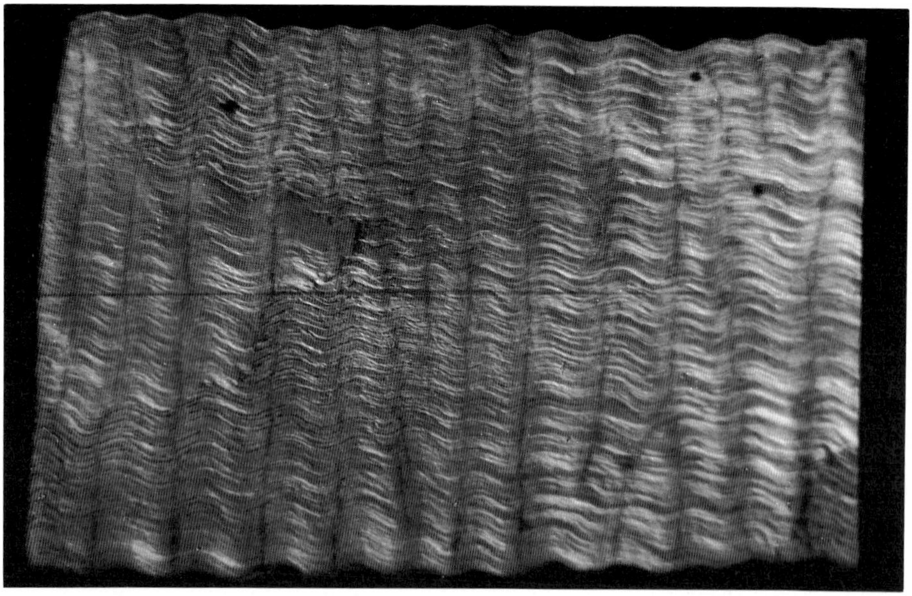

Fig. 5.10. Edge dislocations producing a pifurcate pattern in wavy grain of *Fraxinus excelsior*. (Zagórska-Marek 1985)

consistent and continuous undulations on a split radial face being maintained over many years' growth. Occasionally such undulations contain equally persistent dislocations. Zagórska-Marek (1985) described very short domains (measuring 2–3 mm longitudinally) resulting in very tight wavy grain in *Fraxinus excelsior.* At some locations on the cambial surface transverse bifurcations of domains occurred, which she likened to edge dislocations in the molecular networks of crystals. The point of the bifurcation did not shift horizontally, but moved longitudinally along the stem as part of the total domain pattern. For the period studied (ca. 30 years) particular dislocations did not show significant changes in their dimensions or configurations, but left behind a consistent bifurcated pattern in the wavy grain (Fig. 5.10). Zagórska-Marek suggested that dislocations and their persistence could be important in understanding the nature of morphogenetic fields.

5.5 Wave Aspects of Domain Structure and Function

Hejnowicz (1972) set forth those features of cambial domains that invite comparison with wave phenomena:

1. Domain patterns containing spatially separated alternating orientational tendencies, manifested by S and Z morphogenetic events, migrate along the cambium creating grain patterns in wood.
2. As a consequence, the orientational *tendency* of cambial initials changes cyclically as S and Z domains follow in turn.
3. The S–Z orientational *behaviour* of cambial cells is a function of distance along the stem and of time.
4. One can convert from the first function to the second by invoking the rate of movement of a domain pattern: the phase is equal to the length of domain divided by the rate of migration.
5. In all of these respects (a phenomenon with periodic reversal of direction, as functions of time and distance) the domain pattern has the attributes of a wave system.

Each moving wave is characterised by its wavelength λ, period T, and velocity v, which bear the relationship $\lambda = vT$. In morphogenetic waves in the cambium, velocity is the velocity of domain migration, period is the time between successive identical reversals (e.g., S to Z, or Z to S) at a particular site, and wavelength is given by the length of a pair of domains along the direction of migration (Hejnowicz 1975).

When the frequency of morphogenetic events in the cambium is high, and therefore determines the grain angles that develop, the wave period is represented in the xylem by the *radial distance* between the same phase of neighbouring undulations, and can be converted to time in years by counting the number of annual growth layers within that distance. Wavelength is represented by the *longitudinal distance* between the same phase in two neighbouring undulations.

Hejnowicz (1974) was careful to point out that the morphogenetic wave is a manifestation of unknown processes within the cambial protoplasm. The sup-

Table 5.2. Relationship between the hypothetical morphogenetic wave in the cambium, domain pattern, and grain angle. (Zagórska-Marek and Hejnowicz 1980)

Morphogenetic wave	Domain pattern	Grain
A (amplitude)	Frequency of cellular events	Maximum Z or S deviation of grain (on tangential surface of wood) from direction of the stem axis
λ (length)	Domain length $= \lambda/2$	Height of areas (on tangential surface of wood) with Z or S grain inclination
v (velocity)	Velocity of pattern migration	Inclination of wave front to wood rays (on radial wood surface)
T (period)	Duration in time of two successive domains at a given point in the cambium	Radial distance between two successive wave fronts (on radial wood surface)

posed relationships between these processes and domain patterns and spiral grain are set out in Table 5.2. It may be inferred that the process itself occurs as a wave $\psi(t, z)$ characterised by the known λ and v, but the shape of the wave (i.e., the function ψ) is not known. Only two classes of orientation, S or Z, can be distinguished at the level of morphogenetic events, but this does not imply that there are only two states in the underlying process.

Hejnowicz (1974) suggested the probability that something in the process changes gradually with time and distance. For a working hypothesis he considered the wave as a simple harmonic function of time (t) and distance (z):

$$\psi(t, z) = A \cos(\omega t - kz) ,$$

where ω is the angular frequency, $\omega = \frac{2}{T}$, k is the wave number, $k = \frac{2\Pi}{\lambda}$, and A is the amplitude.

For convenience, it is assumed that the orientation of morphogenetic events (either S or Z) is determined by the sign of this function, as is illustrated diagrammatically in Fig. 5.11.

That the biological rhythms are time-dependent was shown by Krawczyszyn and Romberger (1975, 1980) in their studies of *Nyssa sylvatica* and *Liquidambar styraciflua*. Choosing pairs of stems having common root systems, but having very different diameters, they were able to show that cycles of grain inclination within the stem pairs had a closer relation to calendar time than they had to the amount of radial growth of each stem.

A cambial morphogenetic wave is unusual in that its period is extremely long, averaging about 10 years. In fact, period is the least variable parameter in the waves of different cambia and different species. Hejnowicz (1975) gave as the range of observations known to him, $T = 8-20$ years, $\lambda = 3-1000$ mm, $v = 0.1-200$ mm year^{-1}. Some of these figures have been extended over the years. Wavelengths as long as 7–10 m have been recorded in *Nyssa sylvatica* (Krawczyszyn and Romberger 1980), and in *Entandrophragma cylindricum*

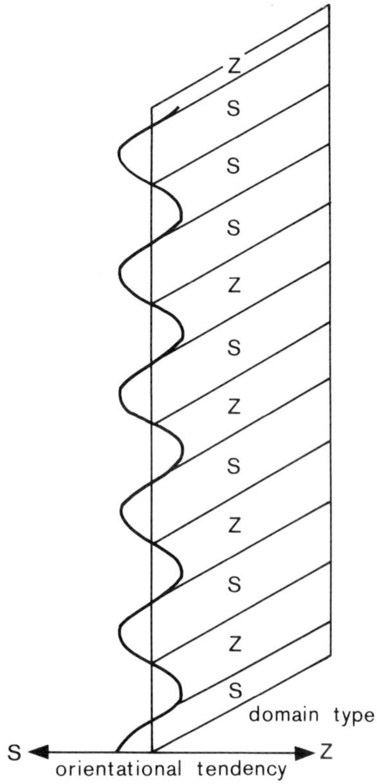

Fig. 5.11. Model of sinusoid wave of orientational behaviour and the corresponding domain pattern in the cambium. The domain pattern is constructed on the assumption that domain type is determined by the sign of the wave. (Hejnowicz 1974)

wavelengths often extended beyond the 3-m boards used in her investigations by Zagórska-Marek (1977). Détienne (1979) observed that a total S–Z wave is completed annually in interlocked grain in *Daniella* spp., indicating T = 1 year (and presumably v = tree height per year). Other observations of relatively short values for T include those of Zagórska-Marek (1984a) in *Tilia*, in which eight grain cycles were completed in 16 years (T = 2 years). Period is less easily determined in tropical timbers such as *Entandrophragma* because annual rings are difficult to define, and it is almost certain that several tangential parenchyma bands are produced annually. However, Zagórska-Marek (personal communication) doubted that T was less than 2 years even for the narrowest pairs of stripes produced by interlocked grain used in her studies. So although the limits of T may now have been extended to 1–20 years, the original proposition still holds that T is the least variable of the cambial morphogenetic wave parameters.

Simple wave patterns in wood cover a very wide range of wavelengths, from the very short, nearly horizontal undulations as seen on a radial face, which subtend only a small angle to the wood rays, up to the nearly longitudinal stripes of interlocked grain which subtend nearly 90° to the wood rays and only a small angle to the annual growth-layer boundaries (cf. Fig. 1.5). Since period is the least variable wave parameter, it follows that within broad limits $\lambda \propto v$ for all species, and the longer the domain the faster it will migrate along the stem (Table 5.3).

Table 5.3. Relation between the lengths of domains and velocity of their migration along the cambium. (Hejnowicz 1973a)

Species	Domain length (mm)	Migration per 10 mm of xylem laid down (mm)	Radial interval between reversal of domain type (mm)
Fraxinus excelsior	2.5	0.8	31
Acer pseudoplatanus	2.3	0.6	38
Betula verrucosa	7.0	4.8	15
Platanus occidentalis	80 – 280	65 – 120	17
Picea abies	7.0	7.0	10
Guibortia ehie			
Short domains	7.5	4.6	20
Long domains	1000	600	20

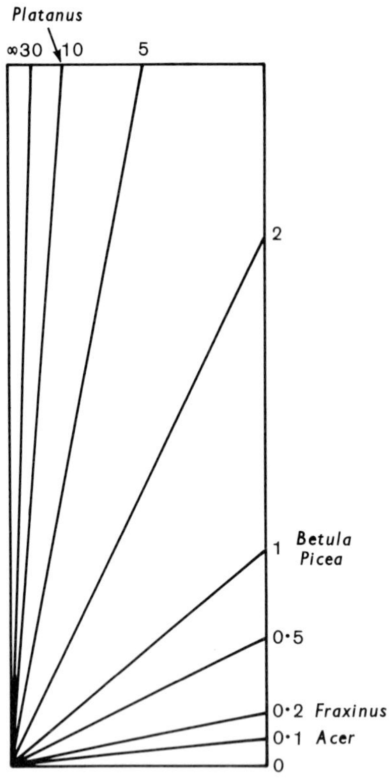

Fig. 5.12. Relation between the velocity of wave movement in the cambium and the inclination of undulations on the radial surface of wood with wavy and interlocked grain. *Numbers* show velocity of domain movement in millimetres per millimetre of xylem produced. (Krawczyszyn 1972)

The velocity of domain movement (expressed as mm mm^{-1} of radial growth) need only vary from 0.1 to 10 to increase the angle between undulations on the radial face and the wood rays by approximately 80° (Fig. 5.12).

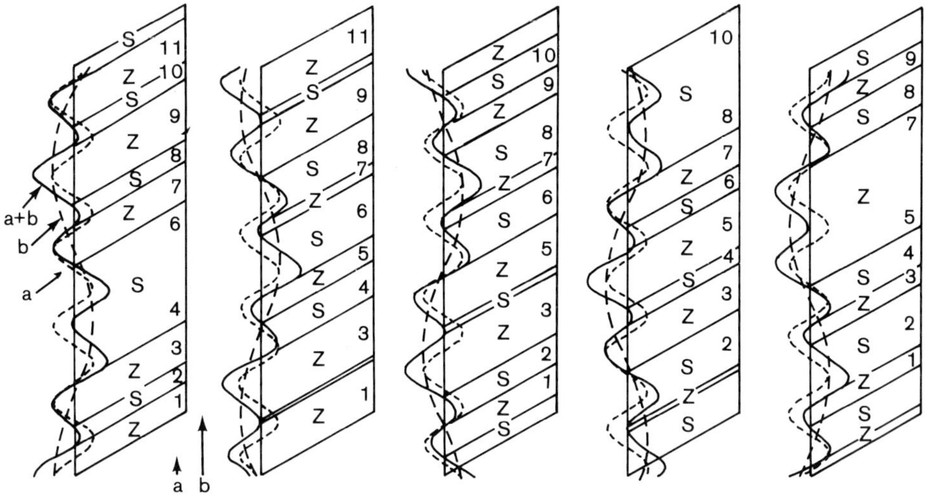

Fig. 5.13. Diagram of two waves (*dashed lines*) superposed, the resultant wave (*continuous line*), and the corresponding domain pattern in the cambium, at five equidistant times. During successive stages in the figure, the component waves are displaced by the vectors *a* and *b* as indicated by the *arrows* of different length. The vectors are chosen so that the quotient that gives the period of the wave is the same for both waves; $(\lambda a)/(va) = (\lambda b)/(vb)$. The domains are *numbered* to facilitate their identification. Fluctuation of domain length (pulsation) can be seen. (Hejnowicz 1974)

5.5.1 Wave Interactions

So far only simple waves have been considered, but one of the most convincing pieces of evidence in support of the theory of morphogenetic waves must be that, where more than one wave system is present, they can be seen to interact in a way that is characteristic of other wave systems.

Hejnowicz (1974) considered the case where two waves of different wavelenghts λ_a and λ_b, and different velocities v_a and v_b, but of similar period, are present in the same cambium. Figure 5.13 illustrates the unidirectional model of the interference of two such waves, and the corresponding domain pattern, in successive time periods. A result of wave interaction is that domains are of different length at any one time, and that each domain varies in length with time (pulsates), even to the extent of disappearing temporarily.

Hejnowicz then pursued this matter further, using two radially sawn boards of *Fraxinus excelsior* showing complex wavy grain, and cut from separate trees, to test this model. The wood in both boards had close wavy grain ($\lambda = 5.5$ and 7.0 mm) with angles of 3.5° and 7.0° respectively to the wood rays. However, when the boards were cut into radial strips 30 mm deep, and each strip was split to show the radial longitudinal face, it became apparent, when the total grain pattern was reconstructed, that there were superimposed long wave patterns in each board. These long waves had $\lambda = 45$ and 31 cm respectively, which made an angle of about 80° to the wood rays in each case.

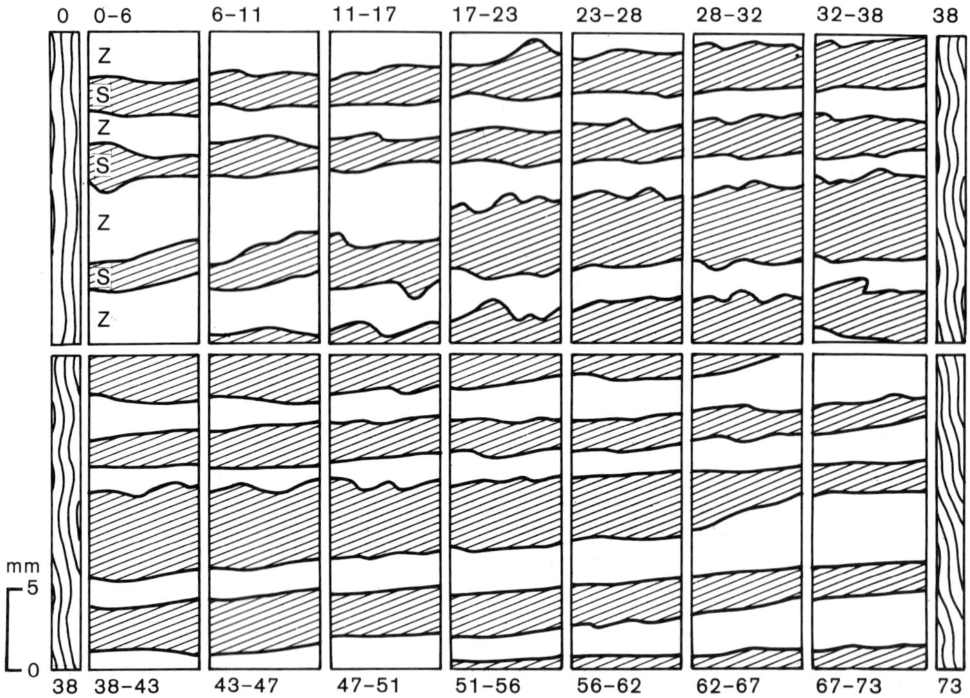

Fig. 5.14. Domain pattern in *Fraxinus excelsior* at radial intervals (mm) indicated by the *numbers* above each map. Close wavy grain ($\lambda = 7$ mm) is produced by domain movement upwards. This is superimposed on a long wave pattern ($\lambda = 30$ cm) resulting from pulsation of domain height. Overall LH grain angle at 73 mm (*bottom right*) represents a small part of the long wave, and is the result of the dominance of deep S domains between 17 and 52 mm. In the original diagram about 60 symbols were shown in an average domain, indicating the locations of S or Z morphogenetic events. (After Hejnowicz 1974)

Blocks used for microscopic examination did not cover the entire wavelength of the longer waves, but covered one phase with parts of the other phase at either end of the sequence. By examining sections cut from the terminal xylem layer in each annual ring, Hejnowicz constructed a series of 14 domain maps, one for each annual period at the same site in the cambium (Fig. 5.14). As usual, S and Z domains proved to be the basis for the development of the close undulations in the grain, as could be seen by comparing domain activities with the changing grain angles of the narrow waves. Again, the migration rate was in agreement with the angle between radial undulations and wood rays.

The longer waves, on the other hand, could be related to pulsations in domain length. Thus in one section, where Z domains were longer than S domains, so that Z events were predominant overall, the average grain direction tended towards RH, even though the short wave pattern was maintained. This situation then gradually changed over the same area of cambium, as the Z domains decreased in length and the S domains increased. As the S domains became predominant, so the overall grain direction to the right began to decrease and eventually became directed towards LH, but still with undulations due to the narrow domains.

The whole pattern of pulsating domain lengths not only accounted for the longer waves in the wood, but agreed in all respects with the model illustrated in Fig. 5.13, with each domain varying in length with time, and sometimes nearly disappearing altogether.

Many other examples have been discovered in which more than one wave form exists in a single piece of wood. Pyszyński (1977a) found long ($\lambda = 500$ mm) and short ($\lambda = 6$ mm) waves superimposed on spiral grain in *Aesculus*. The short waves made an angle of 4° to the wood rays, and the long ones 85°.

Krawczyszyn (1977) observed that the normally interlocked grain of *Platanus acerifolia* became straight in the region of the root collar, a zone about 20 cm long in which roots join the main trunk of the tree. Long, rapidly moving cambial domains were responsible for interlocked grain in the tap root and main lateral roots, as they were in the stem, but in the root collar, where cambial domains from all these sources came together and interacted, domains were irregular in shape, size, and duration. As a consequence any trend towards establishing a regular pattern of spirality in this zone was eliminated.

Zagórska-Marek (1977) observed disturbances in the rhythmic waves produced by split radial surfaces of *Entandrophragma cylindricum*, which she attributed to the superposition of a short wave on the very long wave that is the basis of interlocked grain in this species. In a later paper Zagórska-Marek and Hejnowicz (1980) studied small localised wave patterns on the radial faces of boards of *Entandrophragma cylindricum*. These appeared in two forms. In one, short slanting waves crossed the main pattern of near-vertical stripe or ribbon figure (the pattern of interlocked grain). Close examination of these small undulations showed that they commenced as the interlocked grain changed from S to Z inclination, and disappeared (i.e., wave amplitude became zero) on completion of one full cycle of interlocked grain. In this process the short wave moved upwards along the board by half a wavelength. Then, after disappearing momentarily, the short wave was re-established in the same relative position as before, half a wavelength back from its point of disappearance.

The second type of wave left horizontal undulations on the radial face. Each wave was entirely contained within either the dark or the light stripe (T/2) of the interlocked grain figure. That is, wave amplitude commenced and finished at zero, as Z and S phases succeeded each other. Grain inclination in each short wave was in the direction opposite to that of the prevailing stripe. The waves could therefore be explained as a temporary change of phase by T/2 in the oscillation process associated with formation of the long-wave interlocked grain.

This appearance of short waves superimposed on a grain pattern derived from long, fast-moving waves led Zagórska-Marek (1980) to ask whether all complex wave patterns in wood had to be viewed as separate waveforms that produced interference patterns, or whether the mode of wave propagation might, of itself, produce complex figures in some instances. She considered four cases which are illustrated in Fig. 5.15.

A. A morphogenetic wave system is initiated simultaneously over a broad area of cambium, and waves move upwards. Wave amplitude (as indicated by lengths of arrows) remains constant. Result: wavy grain with periodic straightening, as seen in the short waves of *Entandrophragma*.

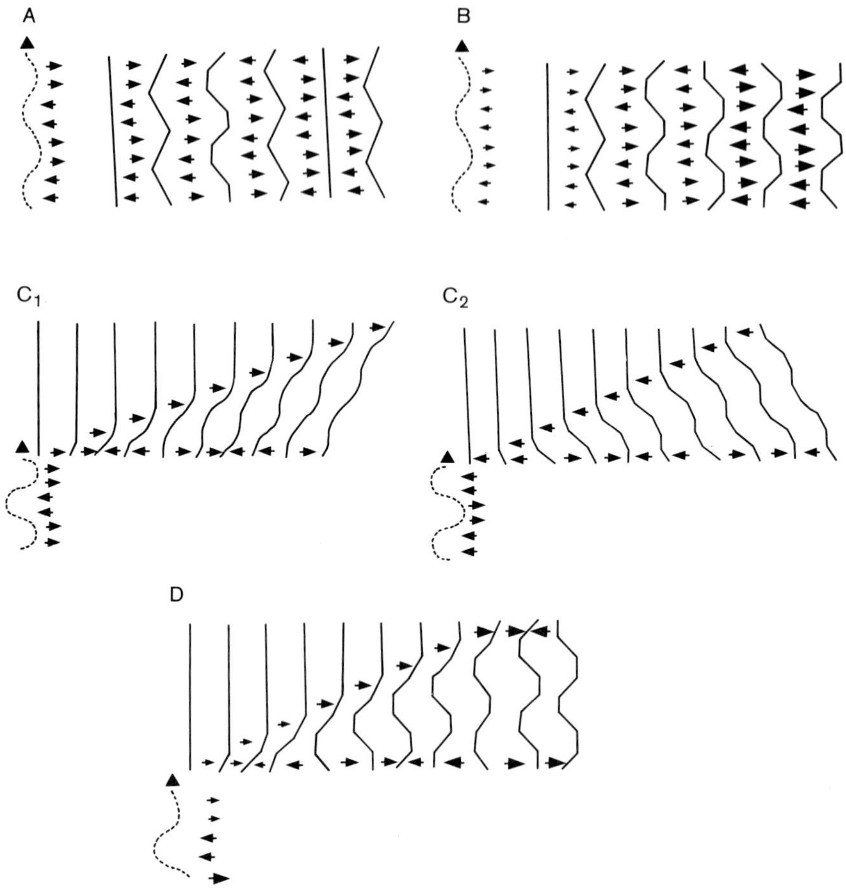

Fig. 5.15 A–D. Effect of the mode of wave propagation on resultant wood grain figure. **A** Morphogenetic wave system initiated simultaneously over an area. Wave amplitude remains constant. **B** Wave system initiated simultaneously over an area. Wave amplitude increases for a time, then remains constant. $\mathbf{C_1}$ Wave system initiated progressively over an area. Wave amplitude remains constant. Initial change of cell direction to the *left*. $\mathbf{C_2}$ As for C_1, but initial change of cell direction to the *right*. **D** Wave system initiated progressively over an area. Wave amplitude increases for a time, then remains constant. (Zagórska-Marek 1980)

B. A wave system is initiated simultaneously over a broad area of cambium, and waves move upwards. Wave amplitude increases for a period after the wave has been initiated. Result: wavy grain which does not show periodic straightening, i.e., "typical" wavy grain (Fig. 2.12).

C. A wave system is initiated progressively from the base of a stem and moves upwards. Wave amplitude remains constant. Result: spiral grain develops, more or less wavy depending on wave amplitude, and tending to left or to right depending on the first change in cell orientation (Fig. 5.15, C1 or C2). Mean spirality is half of the maximum amplitude. Wavy spiral grain has been recorded in *Aesculus* (Pyszyński 1972) and *Platanus* (Zagórska-Marek, personal communication) as illustrated in Fig. 5.16.

Note: This pattern could also be derived from downward movement of the wave system.

D. A wave system is initiated progressively from the base of a stem and moves upwards. Wave amplitude increases for a period after the wave has been initiated. Result: wavy grain that is indistinguishable from B.

It may be noted that, according to this hypothesis, it is necessary for wave amplitude to increase for a time after the morphogenetic wave system has been initiated if normal, continuous, wavy grain is to be established. This applies whether the wave system is initiated simultaneously over a broad area of cambium or is initiated progressively from the base of a stem and moves upwards. Another feature is that morphogenetic waves of very small amplitude would, under C, gradually induce spiral grain with imperceptible waviness.

5.5.2 Complex Wave Interactions

More complex grain patterns have been attributed to wave interactions of the third or higher order (Hejnowicz 1973b). The fact that different wave patterns in the same stem have been shown to have the same period (e.g., in *Fraxinus* and *Entandrophragma*) draws attention to the possibility that such waves could interact to produce a stationary pattern of beats. Charts of interlocked grain prepared by Martley (1920) indicate some system of beats in *Chloroxylon swietenia, Gmelina arborea* and *Hardwickia binata.* The requirement for this development is that the difference between the two wavelengths is small in comparison with the mean value of wavelength.

A model of stationary beats developed in this way is illustrated in Fig. 5.17, and Hejnowicz (1973b) provided the following mathematical explanation:

Let waves ψ_1 and ψ_2 be sinusoid ones, i.e., $\psi_1(x,t) = A\cos(\omega_1 t - k_1 x)$ and $\psi_2(x,t) = A\cos(\omega_2 t - k_2 x)$, where A = amplitude; ω = angular frequency, $\omega = 2\Pi/T$ and k = wave number, $k = \omega/v$.

Applying trigonometry to the sum of waves provides the known relationship:

$$\psi_1(x,t) + \psi_2(x,t) = 2A\cos\tfrac{1}{2}[(\Delta\omega)t - (\Delta k)x]\cos(\bar{\omega}t - \bar{k}x) ,$$

where $\Delta\omega = \omega_2 - \omega_1$, $\Delta k = k_2 - k_1$, $\bar{\omega} = (\omega_1 + \omega_2)/2$, and $\bar{k} = (k_1 + k_2)/2$.

The factor $2A\cos\frac{1}{2}[(\Delta\omega)t - (\Delta k)x]$ plays the role of modulated amplitude, A_{mod}. Since $k_2 \geqslant k_1$ thc curves $\pm A_{mod}$ serve as an envelope within which the oscillations of the factor $(\bar{\omega}t - \bar{k}x)$ are inscribed. The zeros of cyclically (along the x-axis) changing width of the envelope divide the resultant wave cos $(\bar{\omega}t - \bar{k}x)$ into groups (beats). If $\Delta\omega \neq 0$, then the envelope itself represents a moving wave of second order propagated with constant velocity, $v_g = \Delta\omega/\Delta k$, known as group velocity. If $\Delta\omega = 0$, i.e., $\omega_1 = \omega_2$ (which is true for waves in the cambium), then the time-dependent factor in A_{mod} vanishes. A_{mod} is then given by:

$$A_{mod} = 2A\cos(\Delta k/2)x ,$$

i.e., it changes cyclically along x but is stationary in space ($v_g = 0$).

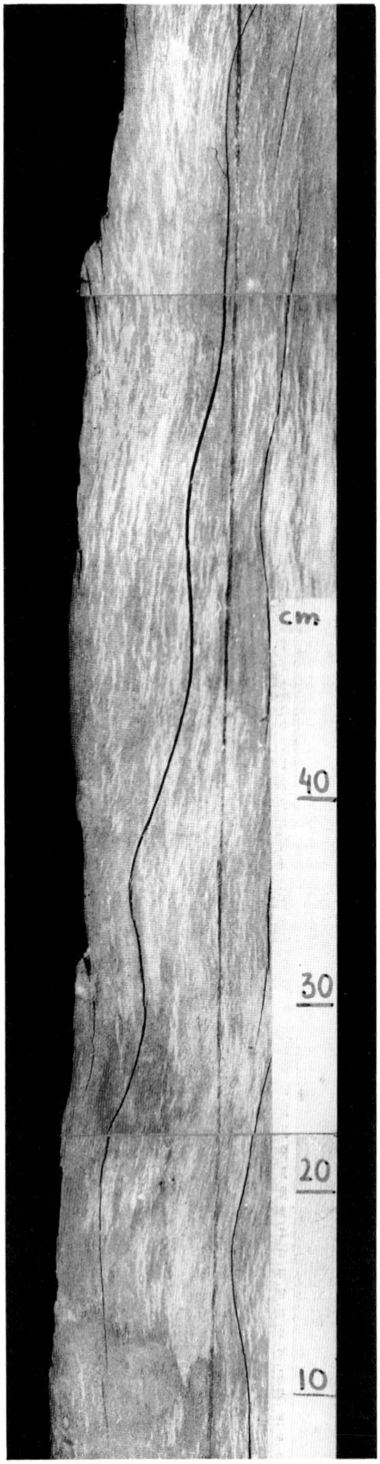

Fig. 5.16. Wavy spiral grain in a log of *Platanus occidentalis*. (Photo B. Zagórska-Marek)

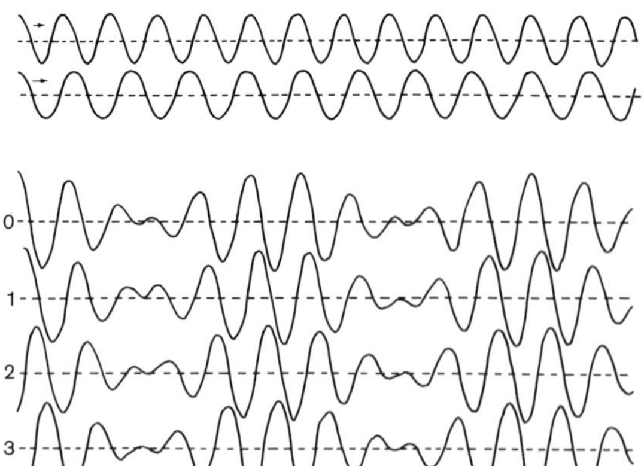

Fig. 5.17. Diagram to illustrate how stationary beats arise when two waves (*upper part*) are superimposed. The waves are of the same frequency but of slightly different wavelengths, and their velocities are indicated by the lengths of the *arrows*. The resultant wave is illustrated at four equidistant times. (Hejnowicz 1973b)

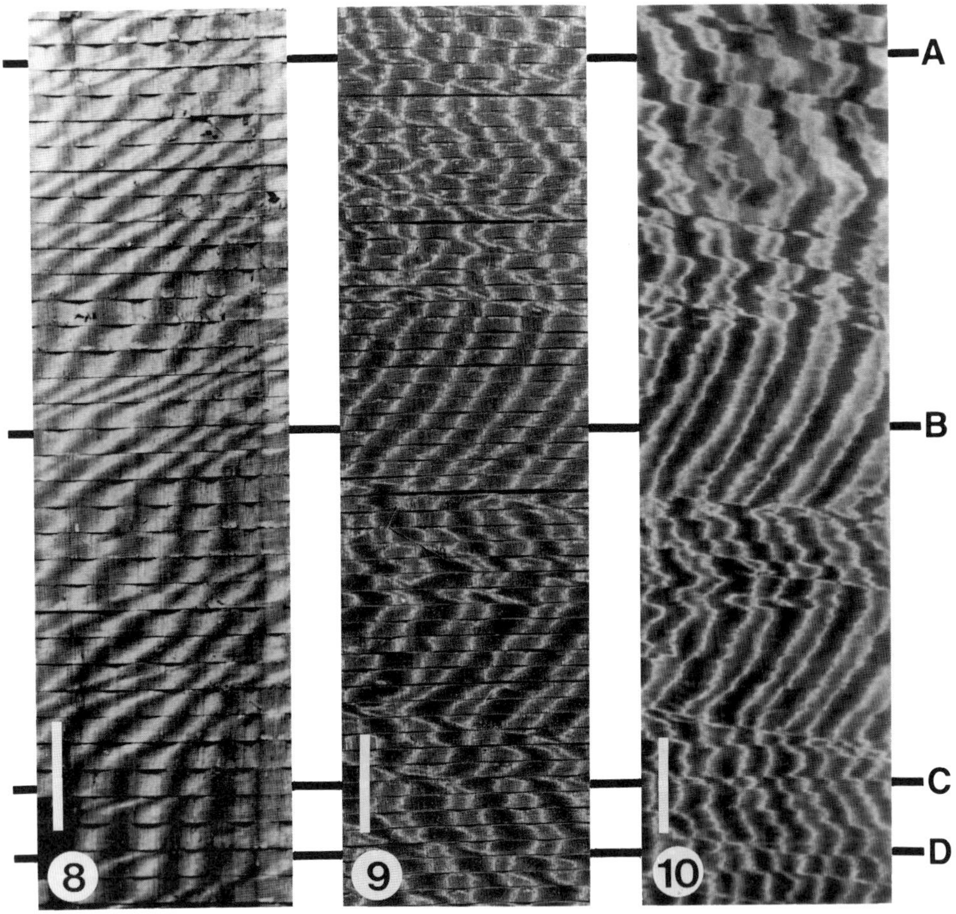

Fig. 5.18. Complex wavy figures visible on radial faces of *Guibourtia ehie*. **8** and **9** show undulations revealed by cutting a radial board into strips and then splitting each in the radial-longitudinal direction. The split strips were then arranged in their original orientation and sequence. **10** is the same board as in **9** but with its planed surface photographed before cutting into strips. The cambium would lie to the *right* of each figure; *white bar* represents 10 cm. (Hejnowicz and Romberger 1979)

Hejnowicz explained that the stationary pattern of beats produced by superposition of waves with equal T but different v will be obtained by waves of any other form, and is independent of the sinusoid waves used in this particular example.

Hejnowicz (1973b) then went on to consider a complex grain pattern in *Guibourtia ehie* (Fig. 5.18, board 8). This example is also included by Hejnowicz and Romberger (1979). It is interpreted as resulting from superposition of a least three waves ψ_1, ψ_2, and ψ_3, of equal T, the wavelengths of which are different such that $\lambda_1 > \lambda_2 > \lambda_3$. ψ_1 and ψ_2 then produce by superposition wave ψ_{12} of $\lambda_{12} = (\lambda_1 + \lambda_2)/2$, the amplitude of which disappears in some regions due to a beating phenomenon. In these regions the third, shorter wave ψ_3 becomes manifest.

Table 5.4. Relation between type of wood grain and resultant wood figure. (Hejnowicz and Romberger 1979)

Type of grain	Type of figure[a]
Straight	None
Spiral	None obvious
Interlocked	Longitudinally banded
Wavy	Wavy, curly
Short wavelength	Fiddleback, tiger
Medium wavelength	Oblique wavy
Long wavelength	Steep wavy (grading into interlocked)
Interlocked wavy, medium (superimposed)	Moiré
Interlocked wavy, long (superimposed)	Chequered
Irregular	Irregular (flamed, drapery, floral, etc.)

[a] As seen on radial surfaces. Tangential or oblique surfaces reveal modified or different aspects of these figures.

A second complex wave pattern in *Guibourtia ehie* (Fig. 5.18, board 9) was interpreted by Hejnowicz and Romberger (1979) as being the result of nearly "pure" waves of the same length but of opposite direction. This is most clearly seen at levels B and C. Moiré, quilted, or chequered patterns can be a manifestation of the standing wave resulting from superposition of two waves, of the same length and amplitude, moving in opposite directions, and some of these effects are seen at levels A and D in Fig. 5.18. Hence the claim in the title of the paper by Hejnowicz and Romberger, "The common basis of wood grain figures is the systematically changing orientation of cambial fusiform initials". Table 5.4 summarises the different types of figures arising in wood with various grain configurations.

5.6 Anatomy of Slow Changes

After considering wavy, interlocked, and complex wave patterns, it may seem strange (not to say anticlimactic) to come back to "simple" spiral grain. However, the normally slow development of spiral grain (i.e., when compared with the more rapid fluctuations of other patterns) is paralleled by a relatively low frequency of morphogenetic events. For this reason the anatomical basis of spiral grain development is, to some extent, less clear than that of the more complex grain patterns.

Bannan (1964b) concluded that spiral grain does not appear to be due to a simple cause. He acknowledged the importance of conformity in the orientation of anticlinal divisions and the direction of cell elongation, but felt that other features such as cell loss and domain reversal would act as controls to restrict spiral development. His conclusion was that excessive spirality arises when the controls are rendered inoperative over a major part of the cambium.

The selection of one group of morphogenetic events merely because they leave behind the clearest record in the xylem is perhaps unfortunate when they are used to examine slow changes, as in spiral grain. In his own studies of *Fraxinus ex-*

celsior, Hejnowicz (1972) was forced to rely mainly on ray-splitting and ray-uniting events to compare samples of straight and spiral grain, and he admitted that the frequency of these particular events was inadequate to define domains or to plot their courses. But even when the full "standard" range of morphogenetic events has been available for examination using serial sections, as in studies of coniferous woods, the adequacy of these events alone to account for observed grain deviations has been questioned (Nicholls 1965, Bannan 1966b, Harris 1973). In other words, angular changes calculated from a certain frequency of pseudotransverse anticlinal division, followed by a known rate of intrusive growth, could not account for the observed changes of grain angle.

Hejnowicz and Romberger (1973) agreed that these observations might be valid if the rate of anticlinal division were no greater than in cambium producing straight-grained wood, but maintained that cell failure occurred in groups, thus minimising its effect on grain angle. In their experience with wood from forest trees, changing grain angles always arose from high rates of anticlinal division. This is not so when sloping grain is induced artificially by stem girdling or creation of cambial bridges (Harris 1973, Savidge and Farrar 1984). Under these conditions processes that are not included in the standard morphogenetic events (including imperfect periclinal division and sideways growth of cell tips) can be seen to contribute to re-orientation of fusiform initials.

Unfortunately, few examples of naturally developed spiral grain have been examined with these other processes in mind, but Savidge and Farrar (1984) found that imperfect periclinal divisions made more important contributions than pseudotransverse anticlinal divisions towards inducing severe natural spiral grain in *Picea glauca*.

If the four morphogenetic events used by Hejnowicz (1972 et seq.) have to be seen as only a part of the total changes that may be involved, this possibility has no bearing on their validity and efficacy for defining domains and explaining rapidly changing grain patterns, where total numbers of events are high. In this the studies initiated by Hejnowicz have been brilliantly successful. On the other hand, the standard morphogenetic events may not be required, by themselves, to account quantitatively for the entire change of angle either during rapid fluctuations or, more especially, during gradual changes.

Hejnowicz himself (1974) wrote that pseudotransverse anticlinal divisions are only a symptom of unknown processes within the cambial protoplasm. Harris (1973) suggested that the direction of pseudotransverse division might be a result of cytoplasmic stress induced by the kind of differential growth that manifests itself in sideways development of cell tips. Indeed, one might speculate that a greater degree of cytoplasmic stress (or the stronger influence of underlying directional forces) is required to induce uniformity in the direction of anticlinal division than to exert a slow, but persistent and effective influence on the other sources of cell re-orientation. If this is so, it could explain some of the apparent anomalies that have arisen in cambial studies, and which have led writers to look for other than anatomical explanations.

Any argument about the means by which fusiform initials re-orientate is, however, less important than what is now clearly recognised: that many of the events leading to changing grain angles are organised and co-ordinated in a way that corresponds to wave patterns moving along the cambium. This has been the greatest advance in understanding changing grain angles in recent years. It has raised the study of grain patterns in wood to a position of significance in plant sciences, and future studies may well provide solutions to many of the problems associated with plant polarity and morphogenesis.

5.7 Conclusions

1. The arrangement and orientation of cambial initial cells very largely predetermines the relative position and arrangement of the various xylem derivatives, and hence the grain angles in wood.
2. Several types of oriented morphogenetic events occur in the cambium. Four types of events that have been shown to leave behind a clear record in the xylem can be used to study the origins of changing grain angles. The events are of two forms: those that tend to change existing orientation towards the right (Z events), and those that induce changes towards the left (S events).
3. The orientation of morphogenetic events within the cambium is generally nonrandom. At any one time considerable areas of cambium, called cambial domains, can be shown to contain morphogenetic events almost exclusively of either the S or the Z type.
4. Within particular areas of cambium, reversals of the predominating direction occur with time. This can be interpreted as migration of domain boundaries along the stem.
5. Domains constitute a phenomenon producing periodic reversals of direction as functions of time and distance, and therefore the domain pattern has the attributes of a wave system, including the essential characteristics of wavelength, period, and velocity.
6. Of these parameters, period is the least variable, ranging from 1 to 20 years. Interlocked grain is created by domains that are very extensive (up to 10 m) in the longitudinal direction, and which move rapidly along the cambium. Tight wavy grain is created by domains that measure only a few millimetres longitudinally and move along the cambium very slowly. Various intermediate dimensions of simple waves are to be found in different species.
7. Multiple wave patterns have been detected within individual cambia. These can interact to produce a great variety of grain patterns including the complex moiré, quilted, and chequered figures.

Chapter 6

Genetics of Spiral Grain

6.1 Basic Requirements

That spiral grain, together with interlocked and wavy grain, are under a considerable degree of genetic control is self-evident. Some species, such as *Pinus roxburghii* are notorious for developing severe spiral grain, whereas others like *Fagus sylvatica* are usually straight-grained. Similarly, whole genera such as *Entandrophragma* and *Eucalyptus* contain many species in which interlocked grain is normal, whereas it is completely absent from other genera. Even wavy grain, though apparently not inevitable in any individual tree, is regularly encountered in some species, whereas there are others in which its presence has never been recorded.

However, it is not such generalised associations that are of interest to foresters and wood scientists, but rather the possibility of influencing the frequency and intensity of these phenomena by selective tree breeding. As yet, few definitive studies have been reported. This is because strictly controlled studies of the heritability of wood properties, and of the potential for genetic gain, have been developed so recently and so rapidly (in common with the science of genetics generally) that it is easy to criticise the seeming superficiality of much of the earlier work. Even today, few genetic experiments have been designed specifically with wood properties in mind. Many of the experiments that have yielded useful results have been set up with quite different objectives. Consequently, it is important to give credit to all genuinely pioneering investigations, at all levels, using their results to sort out the most significant trends, and to achieve better design in future research.

Zobel (1961, 1964, 1965) has indicated the difficulties that may be encountered in assessing heritability of wood properties, in general, and of spiral grain, in particular. Starting with the "immense complexity" of spiral grain formation, Zobel (1965) made it clear that problems in obtaining representative, consistent, and repeatable measurements over a series of experiments have been a primary source of confusion, and one that has prevented more rapid advances.

There is another apparent difficulty which arises in defining what a wood property is for experimental purposes. The basic units of wood are its cells, and the complex series of events that determine their formation and orientation make it seem unlikely at first consideration that any macroscopic measure (which inevitably lumps together the consequences of environmental, physiological (chemical? anatomical?) interactions) would serve as a "unit characteristic" or a trait in its own right, for genetic studies. In one sense maximum efficiency in breeding for a wood property cannot be achieved until it is known how to control the components determining it. Fortunately, such detailed knowledge, whilst obviously desirable, has not proved essential for making an effective start to tree breeding. Considerable advances have been made towards improving even complex properties such as wood density. For practical purposes it therefore makes sense to treat a complex wood property as if it were a simple trait whenever this proves possible, until more detailed knowledge of the components controlling it becomes available.

6.1.1 Concept of Heritability

The most commonly used quantitative measure of inheritance is the ratio called heritability. In essence it expresses the relative importance of genetics and the environment in determining a specific character. In the past, confusion has arisen between the use of broad- and narrow-sense heritabilities. Heritability in the broad sense considers total genetic variability in relation to the phenotypic variability. Thus, if phenotypic variance (V_P) is seen as the sum of genotypic variance (V_G) and variance caused by the environment (V_E) then:

$$\text{Broad-sense heritability } h^2 = \frac{V_G}{V_P} = \frac{V_G}{V_G + V_E} .$$

Heritability in the narrow sense considers only the additive portion of the genetic variability in relation to the phenotypic variability. The additive portion of the genetic variability refers to those effects whereby offspring tend, on the average, to be intermediate between their male and female parents. Where offspring averages tend to depart from mid-parent values non-additive effects may be inferred, which comprise the effects of dominance and epistasis. Since genotypic variance can be regarded as the sum of the additive (V_A) and non-additive components (V_{NA}):

$$\text{Narrow-sense heritability } h^2 = \frac{V_A}{V_P} = \frac{V_A}{V_A + V_{NA} + V_E} .$$

Broad-sense heritabilities are typically estimated using vegetatively propagated groups, or clones. To estimate narrow-sense heritabilities only sexually propagated plants can be used, using either the relative magnitudes of between-family and within-family variance (sib analysis), or regressions of offspring means on parental values.

Consideration of the two formulae for heritability points to two aspects of great importance for sound genetic experimentation and interpretation of the results. An estimate of a genetic variance is a statistic for which reasonably close confidence limits can be ascribed only when it is based on 50 or more degrees of freedom for families. If there are fewer than ten such degrees of freedom, any estimate of genetic variance is virtually meaningless. Especially where ratios of variance components are involved, adequate sample populations are essential for obtaining worthwhile estimates.

The second feature of the heritability formulae is the presence in their denominators of a term for variance due to environmental effects. Heritability can therefore never be regarded as an absolute property for a given species. At best it will be specific to a particular population within a well-designed experiment carried out under clearly stated conditions.

Clear descriptions of the experimental design and degrees of freedom involved are essential for the proper interpretation of results. Where offspring-parent comparisons are possible, it is obviously preferable to use wood with comparable age of formation, and grown, where possible, under comparable environmental conditions. It must also be borne in mind that heritabilities can change with the age of the tree, either increasing (Zobel 1964) or decreasing (Nicholls 1967b) as the tree grows older. Thus, it must be realised that heritability data based on observations of young trees may not necessarily be indicative of realisable heritability at harvesting age.

This very brief review can give only a broad outline of the types of problems to be faced in genetic experiments, and it is beyond the scope of this book to set out in detail all the requirements leading to meaningful estimates of heritability. In any case, estimates of heritability on their own are not very useful. What is needed is some measure of realisable heritability, information on variability, and knowledge of the scope for selecting for the trait in question.

6.1.2 Genetic Gain and Potential for Improvement

A wood property may be under close genetic control, but the potential for genetic improvement also depends on genetic diversity. Unless the "best" genotype is materially different from the "worst" there is nothing for the tree breeder to work on. Gains depend not only on the magnitude of the heritability but also on that of the selection differential, which in turn depends on the intenseness of selection and the phenotypic variance.

$$\text{Expected Gain} = \text{selection differential} \times \text{heritability} = (\bar{x}' - \bar{x})h^2 ,$$

where $\bar{x}'$ and $\bar{x}$ are the means for the select and base populations respectively.

When the selection differential is expressed in standard deviations it is usually referred to as selection intensity (i).

Then

$$\text{Expected Gain } ih^2\sigma_P = i\frac{\sigma_A^2}{\sigma_P^2}\sigma_P = i\frac{\sigma_A^2}{\sigma_P} ,$$

where σ_P^2 is the phenotypic variance, and σ_A^2 is the additive genetic variance.

Most tree-breeding programmes will have to consider the desirability of modifying many features besides those included in the broad category of spiral grain. Only when the frequency and intensity of interlocked or wavy grain are to be increased for decorative purposes are they likely to be given absolute or top priority. Usually, spiral grain will have to compete with tree vigour, stem straightness, branching characteristics, and other desirable features for inclusion in a breeding programme. If these traits vary independently of each other, geometrical increases in the numbers of trees required to achieve high selection differentials in multiplicity of traits are too familiar to need stressing here. Suffice to say that it is often more realistic to consider, for any one trait, the gains that can be made with a selection rate of 1 in 10 rather than 1 in 1000. At this point it may also be noted that there are seldom indications of strong adverse genetic correlations between spiral grain and other traits.

Although it is difficult to extrapolate from data based on observations of young trees to indicate gains at time of harvest, nevertheless a geneticist owes it to the forester to try to indicate whether potential changes achieved by genetic means have any real significance. In this respect much of the work on genetics of spiral grain in conifers is better placed than that on genetics of other wood properties. This is because the initial spiral developed in wood near the pith is of major concern in many conifers. Data from young coniferous trees, showing the maximum spiral angle reached, and the duration of spiral grain before reverting towards zero, can be used most effectively. Other things being equal, this is particularly true for species in which rotations are likely to be short (e.g., *Larix kaempferi, Pinus radiata, P. taeda*) and in which the so-called corewood or juvenile wood represents a significant part of the total merchantable volume production.

6.2 Geographical Variation

Variation in intensity of spiral grain between sites has already been covered in Chapter 4. The extent to which this variation is a direct result of contrasting environmental conditions or may be attributed to the genetic constitution of local provenances is a question that can be resolved only by well-designed research projects. Zobel (1964) suggested that studies of natural variation should employ a nested sampling design so that regional, site, tree-to-tree, and within-tree variations can all be examined at the same time. Provenances can, of course, show very different degrees of spirality relative to each other when transferred from their natural habitat to a new environment (Champion 1945). This is not necessarily an indication of any lack of genetic control of spiral grain, but may simply illustrate the presence of genotype-environment interaction.

The possible existence of distinct provenances with wood properties different from each other must be an important consideration in any breeding programme. Any one provenance seldom contains all the most desirable features, so that the potential for hybridisation to bring together and optimise designated characteristics can be pursued effectively only if the total gene pool is exploited.

In general, the greater the range of sites to which a species has adapted itself, the greater is the likelihood that it will show differences not only between trees growing in its various natural habitats, but also when various provenances are grown together under new environments (Zobel 1964). However, this does not exclude species with limited natural ranges (e.g., *Pinus radiata*, occurring naturally over barely 8000 ha) which, as we shall see later, may show significant provenance differences in development of spiral grain. Equally, there are species with extensive natural ranges in which geographical races do not seem to differ or in which tree-to-tree variation overshadows other sources of variation on all sites. For example, Webb (1967) found that geographical variation in interlocked grain in

Liquidambar styraciflua was non-significant. Differences between stands did exist but these too were less important than tree-to-tree differences within each.

There is some evidence that whereas the magnitude of spiral angles may differ geographically, the broad patterns of spiral grain tend to remain the same within any species. Lowery (1966) found that the distribution of spiral grain patterns (Sect. 2.3.8) in *Pinus contorta, Larix occidentalis, Pseudotsuga menziesii*, and *Picea engelmannii* differed from one another, but were not affected by geographical location over a wide range of sites.

6.2.1 Provenance Trials

When seed lots from distinct geographical locations are planted together on one uniform site, or a series of sites, it is tempting to interpret the results as providing a definitive answer to potential provenance differences. Unfortunately this is not so. As in other forms of genetic experiment, the results may be valid only for the sites chosen for the experiment. It may be hoped that the sites chosen will prove to be representative of considerable forest areas in their particular country, but wherever future sites for afforestation differ from the trial areas there is a possibility that genotype-environment interactions may become evident. This shows up quite clearly in the following example.

6.2.1.1 Provenances of Pinus radiata

The natural range of *Pinus radiata* comprises three discrete forested areas of coastal California, USA – Año Nuevo Point, Monterey Peninsula, and Cambria – and small stands on Guadalupe and Cedros Islands off the coast of Baja California, Mexico.

Nicholls and Eldridge (1980) described variations in some wood and bark characteristics of limited samples from all five natural geographical areas together with three examples of routine Australian planting stock. Trees were growing at five locations near Canberra, Australia, though only on one of these were all five natural populations represented. Soils throughout the area were deficient in phosphorus. In these experiments no significant differences in development of spiral grain were found between any of the natural populations or the Australian seed lots.

Burdon (1980, 1981) reported on more extensive trials of the three mainland populations plus that from Guadalupe grown in New Zealand, together with two New Zealand populations (Kaingaroa and Nelson). Wood specimens from trees aged 7 years after planting were studied in a nominal sample of 2 trees × 2 sites × 2 blocks × 50 wind-pollinated progenies × the six populations. Spiral angle was measured on wood samples containing five annual growth layers, at the second and fifth annual ring for samples from one site, but only at the fifth ring from the other. In this experiment, population differences proved to be very highly significant ($p < 0.001$) on both sites but grain angles showed considerable population-site interaction. Grain angles were definitely greater in the Año Nuevo population,

Table 6.1. Comparison of spiral angle (°) between populations of *Pinus radiata*. (Burdon and Low 1981)

Population	Both sites Ring 5	Site B Ring 5	Site A			
			Ring 5	Ring 2	Average	Difference
Año Nuevo	4.76	4.95	4.56	7.35	5.95	2.79
Monterey	2.86	3.70	2.03	5.24	3.63	3.21
Cambria	3.22	3.90	2.55	4.72	3.63	2.17
Guadalupe	2.54	3.62	1.47	5.20	3.34	3.73
Kaingaroa	4.10	4.97	3.22	4.65	3.94	1.43
Nelson	3.78	4.90	2.66	5.39	4.02	2.73
Approximate SE	–	0.2 – 0.3	0.27	0.21	0.20	0.26

but comparisons among other populations were rather confused due to site interactions and some marked population differences in the radial gradient of spirality from ring 2 to ring 5 (Table 6.1).

It is of interest that the two sites used for this experiment both lay within the same area of plantations (Kaingaroa Forest), only 32 km apart, although one site was selected as being at a slightly lower altitude and markedly warmer than the other. Both were on pumice soils. Had the conclusions from site B, which was examined first, been the only ones available, they would have indicated that spirality was worse ($p<0.001$) in the Año Nuevo, Kaingaroa, and Nelson populations (averages 4.90° to 4.97°) than in Monterey (3.70°), Cambria (3.90°), and Guadalupe (3.62°) (Burdon 1980). How this assessment was changed in subsequent measurements from the second trial area can be seen from Table 6.1.

So here are three sets of results, based on the natural populations of *Pinus radiata*, which are distinctly different for the three trial areas concerned. The major differences are (1) changes in the relative positions of populations with respect to mean spiral angle on different sites and (2) changes in mean angles between sites (mean angles lower on the warmer site A than on site B in the New Zealand experiment).

Differences between populations in the rate of decline from the high (usually near-maximum) angles at ring 2 out to ring 5, also point to the need for total patterns of spirality to be studied.

6.2.1.2 Provenances of Larix kaempferi

Larix kaempferi (Japanese larch) is indigenous to rather restricted areas of Honshu, the central island of Japan, but has been widely planted in the northern island of Hokkaido, and elsewhere in many countries of the temperate zones. Full provenance trials were established in 1955 as part of a worldwide programme, although it had been known for some time that several provenances could be distinguished which showed different growth characteristics and productivity (Toda and Mikami 1976).

Fig. 6.1. Frequency distributions of grain angles for 20 provenances of *Larix kaempferi* grown on two sites in Japan. (Mikami and Nagasaka 1974)

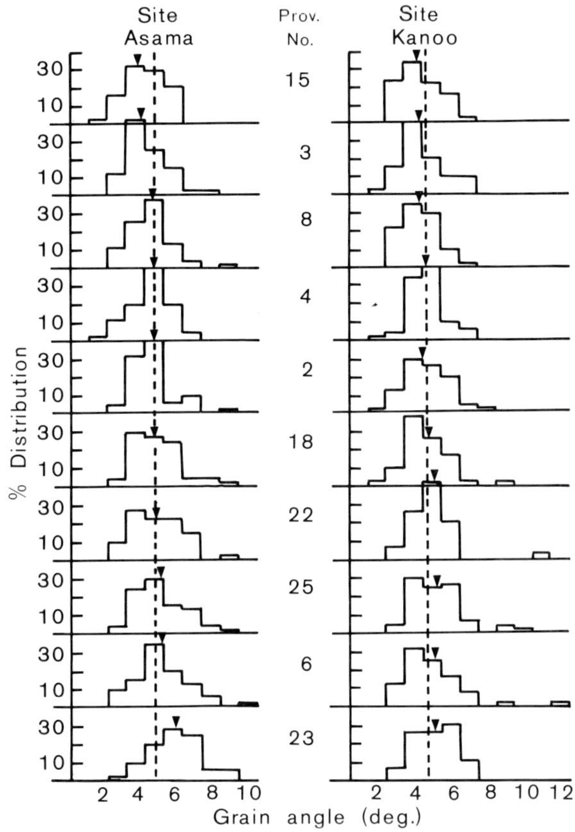

After thinning of the provenance trials in Japan, Mikami and Nagasaka (1974) examined spiral grain in 25 provenances growing in the Asama plantations, and in 10 of these provenances growing in the Kanoo plantations. Significant differences were found between provenances. Smallest angles were observed in provenance No. 5 (Nikoo-Yasyuubara) and the largest were in No. 28 (Mt. Kiso-Komagadake) "which showed extraordinarily high angles". Since warp and twist on drying are a most serious defect in Japanese larch, and one that has repeatedly been related to the severity of spiral grain (e.g., Takizawa et al. 1981), these were seen as being very significant findings.

In this instance, spiral angles produced on the two sites were very similar, and graphs of grain angle distribution within the ten provenances available for study on both sites are illustrated in Fig. 6.1.

6.2.1.3 Provenances of Pinus pinaster

Pinus pinaster is indigenous to the northern Mediterranean (Portugal, Spain, France, and Italy) and also to Corsica and the mountains of Algeria. Nicholls

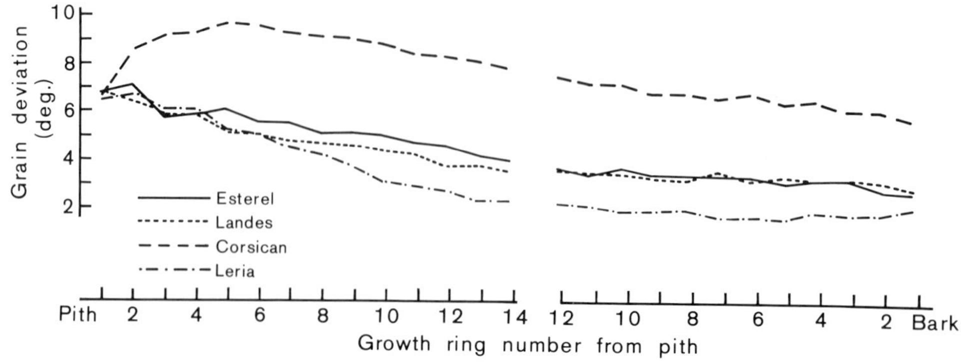

Fig. 6.2. Spiral grain in provenances of *Pinus pinaster* (mean values for ten trees from each provenance). Specimens contained a nominal 26 growth rings, and these have been *numbered* from the pith for the inner 14 rings and from the bark for the outer 12 rings. (Nicholls 1967b)

(1967b) measured spiral grain in ten trees of each of four provenances (Corsica, Esterel, Landes, and Leiria) growing at Somerville in western Australia. With minor exceptions the cross-sectional discs used in this study contained 26 growth layers. The slight discrepancies in numbers of layers were accommodated as shown in Fig. 6.2. The results presented in Fig. 6.2 show distinct differences in development of spiral grain between provenances, with largest grain angles throughout all growth layers in the Corsican provenance, and the smallest grain angles in the Leiria provenance for wood outside the first six rings from the pith.

6.2.2 Use of Provenances in Tree Breeding

The limited number of provenance trials described above serve to confirm that spiral grain is a wood property that may be expected to differ between geographical populations. They also serve to reinforce what have been widely held views for years concerning trees in which spiral grain is an acknowledged problem: that where provenances can be shown to develop spiral grain (or any other major defect) to a serious degree, they should be avoided as a general seed source, and must be treated with caution as a resource for selecting superior trees for breeding. Only if such provenances contain other features of outstanding merit, should they be considered as a source of hybrid production, with the objective of retaining the good features and minimising spiral grain.

6.3 Clonal Variation

The term "clone" refers to a group of propagules that are genetically identical. Such a group normally comprises the original tree (ortet) and the vegetative propagules from it (ramets).

When plants are propagated by asexual means the non-additive gene effects are transmitted in full because the particular gene combinations of individuals remain intact and unchanged. Coppice growth sprouting from a common rootstock provides ready-made clonal groups in many species, and might be expected to reproduce the original plant closely since all aspects of the environment, apart from mutual shading, are guaranteed to be as uniform as possible. Spiral grain in coppice shoots has indeed been observed to follow closely the trends of the parent tree in many species [e.g., *Pinus roxburghii* (Champion 1924) and *Sequoia sempervirens* (Knorr 1932)] though not necessarily with total fidelity to the ortet pattern. However, Northcott (1957) found that sprouts from the same stump of *Alnus rubra* or *Populus trichocarpa* could develop either LH, RH, or zero spirals, which would seem to imply almost total lack of genetic control over spirality in these species.

6.3.1 Clonal Trials

More detailed studies of clonal variation can be achieved by asexual propagation through grafting or rooting of cuttings, and, more recently, micropropagation methods have been used on tree material. However, spiral grain has yet to be studied in material produced by micropropagation methods.

Although clonal material contains the total genotype of the ortet, this does not automatically ensure uniformity in all ramets even when allowance is made for classical environmental effects on the ramets. Cuttings may still reflect the growth phase of the propagation material collected from the ortet. Cuttings or grafts are said to show cyclophysis when they preserve the characters associated with the "physiological age" or maturation state of the shoot from which they originate, and topophysis when they retain the characters peculiar to the position in the tree of that part of the shoot (e.g., branch vs leader) from which they are taken (cf. Nicholls et al. 1977).

Studies of the ortet-ramet relationship in *Pinus radiata* by Nicholls and Brown (1971) and by Nicholls et al. (1974, 1976) were summarised and supplemented with additional data by Nicholls et al. (1977). These studies compared ramets from mature and juvenile ortets with trees grown from seed. (In this species the transition from the juvenile growth phase is normally almost complete by about 7 years, so that "mature" signifies cuttings from trees significantly older than that.) In ramets derived from mature ortets, grain angles in the centre of the stem (rings 2 and 3) were similar to those in the ortet, but the rate of decrease in outer wood was greater in the ortets than in the ramets (Fig. 6.3 a). Differences between trees derived from seedlings and vegetative propagules were inconsistent between growing sites, but where such differences were significant ($p<0.05$) grain angles declined more rapidly outwards along the stem radius in the seedling material.

Where cuttings had been collected from juvenile ortets, differences between ortets, ramets, and seedlings were smaller (Fig. 6.3 b). Except at one site, mean grain angles were slightly larger in the seedlings or ortets than in the ramets, though differences were not statistically significant ($p>0.05$).

Change of grain angle radially across the stem provided the most notable difference between the two groups of ramets. Nicolls et al. (1976) had found that grain angles in the upper stem of *Pinus radiata* were higher in the centre and

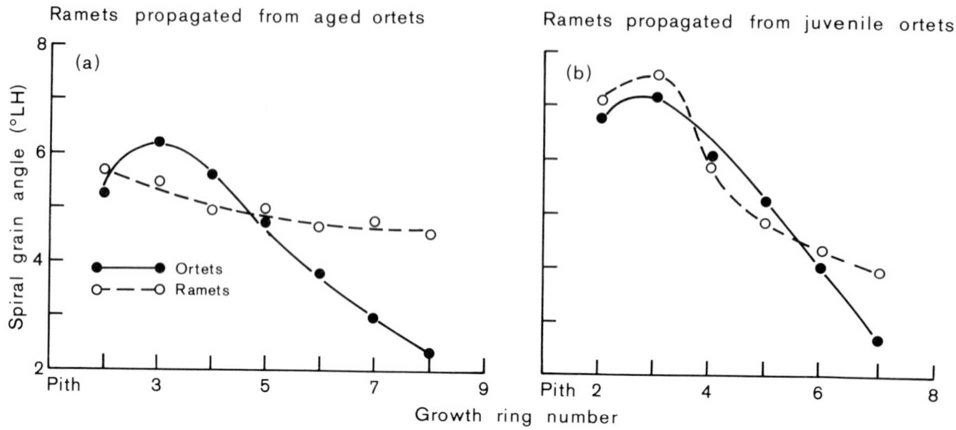

Fig. 6.3. Variation of spiral grain by individual growth layers from pith to bark (**a**) for ramets propagated from mature ortets and (**b**) for ramets propagated from juvenile ortets. (Nicholls et al. 1977)

declined less slowly in outer wood than grain angles in the lower stem, and for this reason Nicholls et al. (1977) suggested that differences between the two sets of ramets illustrated in Fig. 6.3 were compatible with the effects of cyclophysis.

A similar study by Sweet and Harris (1976) was unable to extend observations of spiral grain beyond the fifth annual growth layer, and showed no differences between propagules derived from *Pinus radiata* aged 6, 15, 25, 43, and 66 years compared with trees grown from seed, up to that point. However, marked clonal variation in spiral grain (as in other wood properties examined) confirmed the results obtained by Pawsey (1965), and emphasised the potential importance of including wood properties in clonal selection programmes.

Although the radial pattern of spiral grain in clones may be affected by cyclophysis, there is considerable evidence that the pattern of spirality is under genetic control when the grafted material is strictly comparable. Fielding (1953) illustrated this for two clones of *Pinus radiata*. In one of them, two ramets contained essentially straight grain except for a small zone, a few annual rings wide in the centre of the tree, where grain angles still never exceeded 7°. Three ramets of a second clone, on the other hand, contained severe spiral grain throughout (i.e., out to the tenth annual growth layer) with angles 10° or more in the outer six growth layers.

Harris (1965) reported on spiral grain in five 7-year-old clones of *Pinus radiata* (three trees in each clone). Angles were measured in each annual growth layer in discs cut from every annual height increment. In an appendix to this study, a statistical analysis of spiral grain patterns was developed by W.G. Warren. Orthogonal polynomials for the linear, quadratic, and cubic components were fitted by analysis of variance procedures to the radial patterns of spirality. Results indicated that the pattern of spirality within clones was much more uniform than the pattern between clones.

An outstanding demonstration of distinctive clonal patterns of spirality in grafts of *Larix kaempferi* was provided by Nakagawa (1980). In addition to statis-

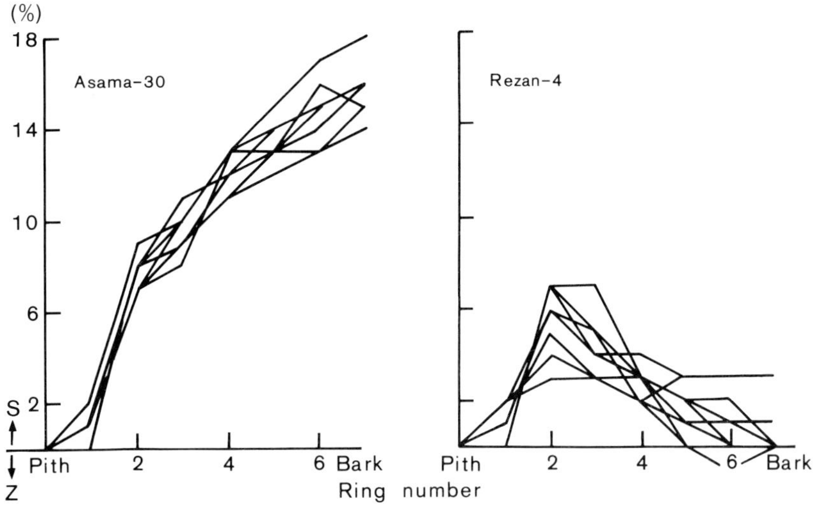

Fig. 6.4. Slope of grain (percentage) measured at breast height in two clones of *Larix kaempferi*. (Nakagawa 1980)

tically significant differences ($p < 0.05$) between clones in both average and maximum grain angles, there was also a significant correlation between maximum grain angle in ortets and ramets. However, it was the remarkable consistency of patterns of spirality within clones that stood out in this study. Two extreme examples of radial changes in spiral angle are illustrated in Fig. 6.4. In clone No. 30 grain angles continue to increase out to the seventh annual growth layer, whereas in clone No. 4 low maximum values are reached by the second growth layer and these revert to nearly straight grain by ring 7.

6.3.2 Broad-Sense Heritability

One of the earliest studies of broad-sense heritability (Dadswell et al. 1961) used rooted cuttings of *Pinus radiata* aged 10 years. Two trees from each of 22 clones were sampled by means of diametric strips cut from standing trees. Data were analysed on the basis of mean values for rings 2–8 and by individual growth layers for rings 6–8. Gross heritability of spiral grain using the mean values for rings 2–8 was 0.66, and for rings 6–8 considered separately was 0.28.

These results led Nicholls (1967a, 1971b) to look for possible changes in heritability with age of wood formation. Three trees from each of 19 clones were examined, again using diametric strips extracted from standing trees. Estimates of broad-sense heritability decreased with increasing distance from the pith (Fig. 6.5). The factors that might be responsible for this trend were discussed in detail by Nicholls (1967a) and included environmental changes within the stand and within the tree, possible changes in gene expression as trees aged, and the increasing relative importance of determination errors in estimating the heritability coefficients. With respect to this last point it may be noted that spiral grain tends to

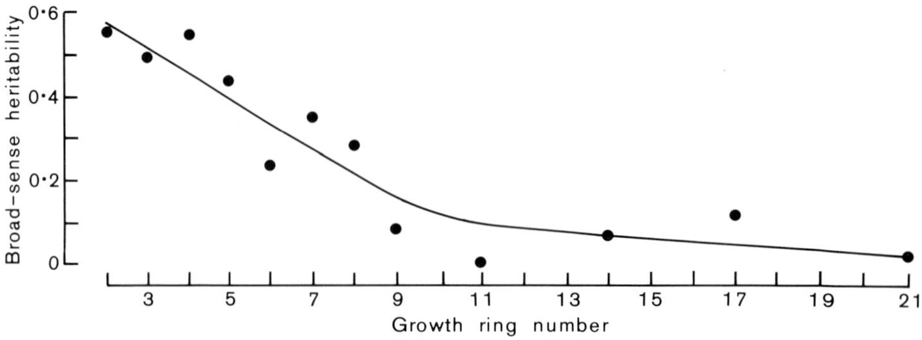

Fig. 6.5. Estimates of broad-sense heritability for spiral grain in individual growth layers of *Pinus radiata*. (Nicholls 1967a)

be at a maximum by ring 2 in *Pinus radiata*, and that by ring 9 values are often approaching zero. Errors of measurement might, therefore, be expected to have more influence on heritability estimates for outer rings than for inner ones. However, the caveat that heritability of any feature need not remain constant throughout the life of a tree is still valid.

Further studies on clones of *Pinus radiata* were reported by Fielding (1967). Table 6.2 is reproduced in its entirety from this publication, because it epitomises the conscientious recording of materials and methods used that are essential for clear understanding of such results. Fielding acknowledged that the large variation of grain angles throughout a tree makes it difficult to provide an index value representative of the stem as a whole. Most of his estimates were based on grain angle measured on the surface of a stem at a particular annual height increment, or on the mean angle of two such observations. In view of the observed within-shoot and tree-to-tree differences in these outer angles, Fielding felt that his estimates of broad-sense heritability tended to be conservative.

Mikami et al. (1972) reported on discs cut from five ramets (grafts) in each of 80 clones of *Larix kaempferi* aged 8–9 years. Mean grain angles for individual clones ranged from 0.6° to 8.1°, and the mean maximum grain angle varied according to clone from 2.4° to 8.6°. Broad-sense heritability based on grain angles for individual rings lay within the range 0.35 to 0.42, and broad-sense heritability based on maximum grain angle within each stem was 0.49.

In later papers (Mikami 1973, Mikami and Nagasaka 1975), measurements of spiral grain in 358 trees on two sites led to the conclusion that correlation coefficients between spiral grain and tree height or tree diameter were low ($r = 0.06-0.10$). It was therefore estimated that tree selection of 1 in 20 for low spirality could be used to reduce the maximum grain angle to 3° and the mean angle to less than 1.3° without adverse effects on growth rate. Furthermore, the fact that most severe spiral angles in *Larix kaempferi* occur early in stem growth, enables tree selection to be made by age 8 or 9 years.

Insofar as the use of coniferous cuttings for afforestation is widespread for species such as *Cryptomeria japonica*, and is being seriously considered for other species including *Pinus radiata*, the gross heritability estimates obtained so far,

Table 6.2. Estimates of the broad-sense heritability of spiral angle in stands of *Pinus radiata* raised from cuttings in the Australian Capital Territory. (Fielding 1967)

Year of estimate	Details	Estimate[a]
1951	Based on 12 clones raised from trees unselected for spiral grain and planted in unreplicated rows in a 6.5-year-old plantation: one measurement of spiral angle made on the surface of the wood at the middle of the internode closest to breast height on 5 – 10 trees in each clone	0.90
1951	Based on 27 clones raised from trees unselected for spiral grain and planted in unreplicated rows in a 12- to 13-year-old plantation: one measurement of spiral angle made on the surface of the wood at the middle of the internode closest to 6 ft. (2 m) above ground on two trees of each clone	0.87
1959	Based on 22 clones raised from trees selected against excessive spiral angle and planted in unreplicated rows in an 11-year-old plantation. Measurement made on wood samples extracted from the trunk at a height of approximately 4 ft. (1.3 m) above the ground (Dadswell et al. 1961)	0.66 (Rings 2 – 8 incl.)
1962	Based on six clones raised from trees unselected for spiral grain and planted in unreplicated rows in an 8-year-old plantation: one measurement of spiral angle made on the surface of the wood at the middle of the longest internode of the fifth annual shoot from the top of the tree and on the north side of the trunk of 10 trees randomly selected in each clone	0.71
1966	Based on three clones raised from trees unselected for spiral grain and planted in a randomised block test with four replications and unit plots of 25 trees in a 6.5-year-old plantation. A minimum of nine trees measured in each plot: one measurement made on the surface of the wood of the fourth and fifth annual shoots from the top of the tree: the estimate based on the mean of the two measurements	0.63*
1966	Based on 12 clones raised from trees selected in a 4-year-old stand for favourable vigour, branching, form and relatively straight grain, and planted in unreplicated rows in a 6-year-old plantation: one measurement made on the surface of the wood at the middle of the basal internodes of the third and fourth annual shoots from the top of five trees selected at random in each clone: the estimate based on the mean of the two measurements. Wombat Creek	0.81
1966	Ditto, the second 1966 estimate above, except that 11 clones were used. Kowen	0.69 (0.83*)

[a] The gross heritability estimates in this table (excepting the estimates marked by *) are expressions of the ratio of genetic to total variance. The estimates marked by an asterisk are based on a total variance estimated in nearby seedlings and the within-clone variance. The number of seedlings was 21 in the first estimate and 34 in the second.

together with the considerable variability recorded, suggest that significant reduction in spiral grain would be possible using quite realistic selection intensities. Presumably the absence of data for broad-sense heritability of spiral grain in hardwoods that are routinely propagated by cuttings (e.g., *Populus* spp.) is an indication that spiral grain is not a significant problem for this group of timbers.

6.4 Sexually Propagated Progeny

Widespread belief in the heritable nature of spiral grain can undoubtedly be attributed in the first place to the observations of *Pinus roxburghii* in India. In contrast to the Indian *Pinus excelsa* in which appreciable spiral grain is rare (Gorrie 1938), the frequency of extreme spirality in *Pinus roxburghii*, and its great variability between sites, led several workers to look for evidence of genetic control (Canning 1915; Smythies 1915, Troup 1916). Results were inconclusive, however, and it was the classic contributions of Champion (1924, 1927a, 1930) that were to have such a profound influence on subsequent attitudes.

Champion collected seeds from parent trees that were either straight-grained or severely spiralled ("twisted" in his terminology), from forest stands that were predominantly straight-grained or in which spiral grain was very frequent. Then he planted the seedlings that he raised on to sites that were themselves notable for producing predominantly straight-grained trees or severely spiralled trees. Despite losses from fire and snow damage, sufficient trees survived through to the sapling stage for Champion (1930) to tabulate his results (here reproduced as Table 6.3), and to publish his conclusions, which, because of their enduring influence, are here reproduced verbatim.

"(i) The chief conclusion which, it is ventured to suggest, is established beyond possible further doubt, is that the production of twisted fibre is a character which is inherited from one generation to the next, independently of any influence which soil and treatment may exert.

(ii) It cannot be asserted that a 100% twisted parent crop will give a 100% twisted second rotation crop from natural regeneration, whether it is protected from adverse external influences or not, but it can be stated already that about two-thirds at least will be twisted beyond serviceability for timber, even given that complete protection which experience has proved impossible in present practice. There are indications that the usual damage from trampling and burning may increase this percentage.

(iii) The use of imported seed from a suitable source will give a crop at least to the sapling stage which is virtually free of twist, even in localities where the original crop is 100% twisted.

(iv) The use of seed from straight trees in localities in which the percentage of twist is moderate or low will similarly give crops in which the percentage of twisted stems will not exceed about 25% and may perhaps be comparable with the parent crop.

(v) The use of seed from straight trees growing among a twisted crop is definitely preferable to that of seed from twisted parent trees growing among a pre-

Table 6.3. Spiral grain in *Pinus roxburghii*. (Champion 1930)

Source of material	A. Average angle of twist at breast height 1930[a]				B. Percentage of stems exceeding 7° twist at breast height			
	Test locality, year of sowings, and twist in surrounding crop				Test locality, year of sowings, and twist in surrounding crop			
	Kaligadh 1921 5% twisted	Maharpali 1917 100% twisted	Khabdoli 1917 10% twisted	Sarna 1919, 1920 60% twisted	Kaligadh 1921 5% twisted	Maharpali 1917 100% twisted	Khabdoli 1917 10% twisted	Sarna 1919, 1921 60% twisted
1. Imported seed from straight trees in straight locality	1 (35)	2.5 (118)	0 (43)	–	0	1	0	–
2. Local seed from straight trees in straight locality	1.5 (30)	6 (63)	–	4.5 (55)	3	25	–	22
3. Seed from straight trees in twisted locality	–	7 (22)	–	9 (134)	–	36	–	45
4. Seed from twisted trees in straight locality	–	11 (22)	–	–	–	59	–	–
5. Seed from twisted trees in twisted locality	6 (32)	11 (63)	9 (20)	–	35	65	60	–
6. Natural regeneration in twisted locality	–	10 (48)	–	7.5 (129)	–	72	–	45

[a] Figures in parentheses under A show number of measurements on which averages were based.

dominantly straight crop, and may result in a new crop about 48% twisted as compared with 60% for the latter.

(vi) Closely comparable results are obtained in three of the four localities demonstrating independence of soil, underlying rock and aspect.

(vii) The results agree throughout with those obtained in 1926 and 1923 and so derived additional reliability particularly where the numbers of trees now measured are small.

(viii) The tentative conclusion is confirmed, which was reached on the data available up to 1921, that left handed twist in *Pinus longifolia* (syn. *roxburghii*) in Kumaon, is of the nature of a Mendelian dominant character which was originated independently in many localities, its extension being due to the long continued selection of the straighter trees for removal."

Although some of Champion's results have been questioned (Kadambi and Dabral 1955, Sen 1955), they have stood, by and large, for more than 50 years as an outstanding contribution to our understanding of genetic control of spiral grain. Indeed, there was a lengthy hiatus between Champion's final report and the next set of observations on sexually propagated progeny in any tree species. Rault and Marsh (1952) observed that certain "strains" of *Pinus roxburghii* (i.e., broadly based geographical origins occurring as commercial seed collections) were more prone to spiral grain than others when grown in South Africa. Furthermore, they suggested that the pronounced skewness of distributions of spiral angle in the more than 300 stands they examined reflected a 'Mendelian ratio' [sic] and therefore provided further evidence of genetic influence.

Several other workers have suggested that the distribution of spiral grain between trees is evidence of genetic ratios. Rao (1954) noted that 70% of *Cedrela toona* were straight-grained, whereas 15% had RH and 15% LH spiral grain. Löhr (1966) observed that 30% of *Aesculus hippocastanum* had RH spiral grain and suggested that this was compatible with the presence of a dominant gene. Kennedy and Elliott (1957) implicated genetic control of spiral grain in *Alnus rubra* on the basis of comparable correlations between radius and spiral angle, and age and spiral angle on two contrasting sites.

An original approach towards assessing heritability of late and early flushing, branch characteristics, and spiral grain in *Fagus sylvatica* and *Quercus robur* was described by Krahl-Urban (1953a, b, 1960, 1961, 1967). By assessing these features in natural regeneration that could be associated with seed parents, he maintained that considerable time could be saved over the lengthy processes of experimental progeny tests. From his observations he concluded that there was strong evidence that parent trees with spiral grain gave rise to a high proportion of spiralled progeny.

6.4.1 Narrow-Sense Heritability

The most extensive work on heritability of spiral grain in sexually propagated progeny has been on *Pinus radiata*. Nicholls et al. (1964) measured spiral grain at the fourth and seventh annual growth layers from the pith in 495 trees grown in the Australian Capital Territory. The experimental plantation contained open-pollinated progenies of 33 trees that had been selected as being typical of plantation-grown *P. radiata*. In each of five block replicates the 33 progeny groups had been randomly allocated to 33 plots or rows, the progeny from each of the 33 parents being confined to a single 23-tree row in each block. Wood specimens in the form of diametric strips extracted from standing trees were taken from the first three trees in every plot.

Variance components were estimated from analysis of variance of the form set out in Table 6.4. In this analysis the environmental variance includes plot-within-block and within-plot components and was therefore defined as $\sigma^2_w + \sigma^2_{bp}$: The genetic component is σ^2_p. Assuming that the relationship between members of a family was half-sib, the estimate of heritability would be given by:

$$h^2 = \frac{4\hat{\sigma}^2_p}{\hat{\sigma}^2_p + \hat{\sigma}^2_w + \hat{\sigma}^2_{bp}},$$

where $4\hat{\sigma}^2_p$ is the estimate of the additive genetic variance, and the denominator is the estimate of the phenotypic variance. However, Nicholls et al. acknowledged the possibility that a proportion of the within-group relationship would be full-sib, and therefore suggested that the additive genetic variance would lie between $2\sigma^2_p$ and $4\sigma^2_p$.

Based on the half-sib relationship, narrow-sense heritability for spiral grain at ring 4 was estimated to be 0.44 ($p<0.01$) and at ring 7 it was 0.24 ($p<0.01$), but if the full-sib relationship applied to part of the progeny these estimates could be reduced to some value between these figures and one-half of them. It was of interest that the reduced heritability between ring 4 and ring 7 paralleled the results for broad-sense heritability given by Dadswell et al. (1961) presented in Section 6.3.2. Harding (1986) recorded similar trends for narrow-sense heritability of spiral grain in *Araucaria cunninghamii*. The similarity between the two sets of values, for narrow- and broad-sense heritabilities in *Pinus radiata*, led Pederick (1971) to suggest that the non-additive component of the genetic variance for spiral grain in this species may be rather small.

Fielding (1967) reported on a progeny test of *P. radiata* in the Australian Capital Territory comprising 14 progenies, using a randomised complete block layout with five block replicates, and plots of 21 trees (three rows of seven trees). Five designated trees were measured in each plot, using slope of grain measured through a small window cut in the bark in the centre of the northern side of the

Table 6.4. Analysis of variance for 495 trees of *Pinus radiata*. (Nicholls et al. 1964)

Source	Degrees of freedom	Expectation of mean squares
Between blocks (b)	4	
Between progeny groups (p)	32	$\sigma^2_w + 3\sigma^2_{bp} + 15\sigma^2_p$
Interaction (b × p)	128	$\sigma^2_w + 3\sigma^2_{bp}$
Within plots (w)	330	σ^2_w

Table 6.5. Analysis of variance for 350 trees of *Pinus radiata*. (Fielding 1967)

Source	Degrees of freedom	Mean squares	F-test	Significance	Variance
Blocks	4	6.00	1.39	n.s.	
Progenies	13	16.88	3.92	$p<0.001$	0.50
Interaction	52	4.30	1.49	$p<0.05$	0.28
Trees within plots	280	2.89			2.89

Additive genetic variance = $4 \times 0.503 = 2.02$.
Narrow-sense heritability = 0.55.

Table 6.6. Analysis of variance for 1043 trees of *Pinus taeda*. (Zobel et al. 1968)

Source	Degrees of freedom	Mean squares
Replicates	2	0.523
Males	17	1.357
Females/male[a]	34	1.566
Pooled error	102	0.243

F-test females/males: $1.5659/0.2431 = 6.44$ ($p<0.01$).
F 0.01 = 1.79, 34, and 102 d.f.

[a] Females within male. Pollinations were by male group, in which two to four females were pollinated by one male. There were no differences between the males (half-sibs) used but large differences between females pollinated with the same male (full-sibs).

longest internode of the sixth annual height increment from the top of the tree (*P. radiata* can form more than one "whorl" of branches annually). Means of values obtained by two observers were used, and the analysis of variance is presented in Table 6.5. Again, it is assumed that the relationship between members of progeny groups is half-sib, but the heritability estimate (for ring 6) is greater than that obtained by Nicholls et al. (1964).

So far no direct comparison of the additive and non-additive components of the genetic variance had been made for any species, but Zobel et al. (1968) were able to do this for *Pinus taeda*. The study was based on measurements of 1043 trees belonging to 52 control-pollinated families in a nested (hierarchical) mating design. Trees were harvested at 4 years of age, and despite their youth considerable differences in grain angle showed up between them. Moreover, spiral grain is essentially a defect of corewood in *P. taeda*, usually, though not invariably, reaching a maximum value near the pith, so that early spiral values are of most significance in this species. Trees from all 52 families were analysed, using a design involving three block replicates times seven trees per family, a total of 21 trees from each family. Grain angle was measured on the outside of each stem relative to the axis of the pith, using a disc cut from approximately 20 cm above ground level. The smallest grain angle for a family of 21 trees averaged 1.7° and the largest 4.8°.

Table 6.7. Variation in spiral angle among *Pinus taeda* families with the same father but different mothers. (Zobel et al. 1968)

Mother	Spiral (°)
	Father No. 48
48 C	2.2
48 D	2.3
48 E	2.6
48 B	3.6
Average	2.7
	Father No. 3
3 C	1.7
3 B	2.7
3 D	3.1
3 E	3.4
Average	2.7
	Father No. 46
46 B	2.2
46 D	2.5
46 C	3.6
46 A	4.2
Average	3.1
	Father No. 10
10 E	4.8
10 C	4.2
10 A	3.4
10 B	2.4
Average	3.7

Analysis of variance (Table 6.6) showed differences among full-sib families to be highly significant, while half-sib families showed no significant differences. The presence of a high proportion of non-additive genetic variance is illustrated in Table 6.7 which gives family means for 4 of the 18 male groups. This illustrates the fairly large differences among progenies with a common male parent. Zobel et al. (1968) concluded that variation of this magnitude and order of frequency among families with one common parent would be expected if appreciable portions of the total genetic variance for spiral grain were non-additive.

The significance of this result is that improvement of a character governed by non-additive genetic variance cannot be achieved by means of a breeding programme using conventional open-pollinated multi-parent seed orchards. Improvement would be possible only by finding out which particular crosses give rise to progeny with smallest grain angles, i.e., with the best specific combining ability in this respect, and by mass production of seed or propagules of the best specific crosses.

Very few of the 4-year-old trees of *P. taeda* had bad spiral grain. Of the 1043 trees examined, only 19 had angles of more than 7°, and only two had spirality as great as 9°. Based on this very young material, Zobel et al. (1968) concluded that spiral grain in corewood of *P. taeda* appeared to have limited significance.

Table 6.8. Variation in spiral grain among families of *Pinus radiata* listed by male and female parents. (Pederick 1971)

Male parent	Female parent	Grain angle (°)	Average for male group
MY7	NZ55	2.23	
	ACT956	2.44	
			2.33
W45	BR48	1.86	
	MY2	4.00	
			2.93
CR54	BR26	2.43	
	BR1	2.83	
	ACT38A	2.84	
	CR40	3.27	
	CA17	3.88	
	CR53	4.60	
			3.31
CR55	ACT112A	3.33	
	CA6	3.33	
	CR56	4.01	
	AS2	4.35	
	CR41	4.58	
	AV21	4.86	
			4.08
CR43	ACT956B	4.11	
	WA2	4.81	
			4.46
CA4	AV7	4.27	
	NZ37	5.00	
	NZ7	5.30	
	AV22	5.37	4.98

Average of all families 3.81 ± 0.07 (528 trees).
Average unselected controls 4.01 ± 0.22 (48 trees).

These results were unexpected in view of the evidence that had been obtained for *P. radiata*, and Pederick (1971) therefore set out to obtain a direct measure of the additive and non-additive components of genetic variance in *P. radiata*. A control-pollinated progeny trial had been planted in 1967 with 1-year-old seedlings of 50 families. From these it was possible to choose 22 families that represented a nested mating design with six "male groups", each containing a common pollen parent but with entirely different mother trees (Table 6.8). Although the design could not be completely balanced, this was no disadvantage in the analysis because additive genetic variance was estimated from the male-group mean square, and the between-females-within-male-groups mean square provided an estimate of additive plus non-additive variance, from which non-additive genetic variance could be calculated by subtraction.

Grain angles were measured through windows cut in the bark on either side of a stem at a height of 30–60 cm above ground level. The great majority of mea-

Table 6.9. Analysis of variance and calculation of narrow-sense heritability for *Pinus radiata*. (Pederick 1971)

Source	Degrees of freedom	Mean Square	F-test	Expection of mean square
Reps	5	35.381		
Between male groups	5	136.953	5.33 **	$\sigma_w^2 + 2\sigma_t^2 + 8\sigma_p^2 + 48\sigma_f^2 + 155.2\sigma_m^2$
Between females within male groups	16	25.678	2.71 **	$\sigma_w^2 + 2\sigma_t^2 + 8\sigma_p^2 + 48\sigma_f^2$
Between plots	105	9.479	1.43 *	$\sigma_w^2 + 2\sigma_t^2 + 8\sigma_p^2$
Within plots	396	6.632	6.23 **	$\sigma_w^2 + 2\sigma_t^2$
Within trees	528	1.064		σ_w^2
Total	1055			

* Differences significant at $p<0.05$; ** differences highly significant at $p<0.01$.
σ_m^2 = Variance due to differences among progenies from different males;
σ_f^2 = variance due to differences among progenies of females mated to the same male;
σ_p^2 = variance due to differences between plots;
σ_t^2 = variance due to differences between trees within plots;
σ_w^2 = variance due to differences within trees.
Components of variance:
σ_m^2 0.7155 ± 0.3350 (SD)
σ_f^2 0.3374 ± 0.1273 (SD)
σ_p^2 0.3559
σ_t^2 2.7838
σ_w^2 *1.0639*
σ_p^2 5.2565
Estimate $\sigma_A^2 = 2(\sigma_m^2 + \sigma_f^2) = 2.1058$;
Heritability (h^2) = $(\sigma_A^2/\sigma_p^2) = 0.40$.

surements would have been made at two rings from the pith (the usual position of maximum spiral angles in this species), but differences in growth rate may have introduced variations up to half a year.

Considerable variations in spiral angle were recorded, not only between families but also between trees within plots. Readings from opposite sides of the stem were often in close agreement, but differences of 3°–4° were not uncommon. A number of trees had straight grain, and the tree with greatest spirality had readings of 10° and 14°.

Evidence for genetic variation in spiral grain can be seen in Table 6.8. Average values for full-sib families varied from 1.9° to 5.4°. The average angle for the 528 experimental trees was slightly lower than that for the unselected controls but the difference was non-significant ($p>0.05$). Analysis of variance (Table 6.9) showed differences among half-sib families (male groups) and among full-sib families (females in male groups) both to be highly significant ($p<0.01$).

The components of variance calculated from the mean squares are also given in Table 6.9. The variance due to differences among progenies from different males provides (from theory) an estimate of one-quarter of the additive component of the genetic variance; and the variance due to differences among progenies of females mated to the same male provides an estimate of one-quarter of the additive plus one-quarter of the non-additive components of the genetic variance. Pederick pointed out that whereas σ_f^2 should exceed σ_m^2 this was not so in these results (a result that had also been recorded elsewhere) and he suggested that the relatively small number of families in this experiment may have introduced a sampling error. Because the parent trees had been selected from widely scattered areas he did not believe that in-breeding could have produced this anomalous result.

Since $\hat{\sigma}_f^2$ is much less than $\hat{\sigma}_m^2$ it was concluded that the non-additive component of the genetic variance is negligible or zero in *P. radiata*. Accordingly $\hat{\sigma}_m^2$ and $\hat{\sigma}_f^2$ could be pooled to provide a better estimate of the additive component of genetic variance:

i.e., $\hat{\sigma}_m^2 + \hat{\sigma}_f^2 = \frac{1}{2}\hat{\sigma}_A^2$

and the estimate of narrow-sense heritability obtained in this way was 0.40.

A selection intensity of one tree in five represents a selection differential of 1.40 standard deviations. The improvement possible would then be $1.40 \times 2.27 \times 0.40 = 1.27°$ and the population average for spiral grain at ring 2 would be reduced from 3.81° to 2.54°.

The most recent estimates of narrow-sense heritability of spiral grain in *P. radiata* are derived from provenance-progeny trials in New Zealand (Burdon 1980, Burdon and Low 1981) that are referred to in Section 6.2.2. Estimated phenotypic coefficients of variation for spirality at ring 5 were 46% for the mainland (Californian) populations, 43% for the two New Zealand populations, and 54% for the Guadalupe population.

It may be recalled that spiral grain at the fifth annual growth layer had been measured in the trials at two sites, but that spiral grain at the second annual growth layer had been measured on only one of these sites. Overall estimates of

Table 6.10. Estimates of variance and heritabilities for spirality in provenance-progeny trials of *Pinus radiata*. Site A with site B values (Burdon 1980) in parentheses. (Burdon and Low 1981)

I. Mainland (Californian) populations

Variable	$\hat{\sigma}_A^2$	$\hat{\sigma}_P^2$	$\hat{h}^2$
Spirality, ring 2	0.97	3.54	0.34
Spirality, ring 5	6.35*** (2.26)	5.12	1.24*** (0.62)
Average angle	2.85**	3.04	0.94**
Difference (2 – 5)	3.76*	5.27	0.71*

II. All populations (four native, two New Zealand)

Variable	$\hat{\sigma}_A^2$	$\hat{\sigma}_P^2$	$\hat{h}^2$	$C\hat{V}_A$ %	$C\hat{V}_P$ %
Spirality, ring 2	2.51	3.73	0.67	29	36
Spirality, ring 5	6.20*** (3.80)	5.36 (3.98)	1.16*** (0.96)	83	83
Average angle	3.29	3.12	1.05***	43	43
Difference (2 – 5)	4.26**	5.29	0.81**	78	87

* Differences significant at $p<0.05$; ** differences highly significant at $p<0.01$; *** differences very highly significant at $p<0.001$.

Table 6.11. Estimates of parameters, population by population, for spiral grain in provenance-progeny trials of *Pinus radiata* on two sites in New Zealand. (Burdon and Low (1981)

Population	Assuming no interaction					Family × site interaction	
	$\hat{\sigma}_f^2$	$\hat{\sigma}_w^2$	$\hat{\sigma}_A^2$	$c\hat{v}_A$ %	$\hat{h}^2$	F-test	$\hat{\sigma}^2$ (approx.)
Año Nuevo	0.807	4.500	3.228	38	0.61*	0.98 NS	0
Monterey	0.559	3.679	4.238	72	0.53*	1.82**	1.22
Cambria	0.536	2.922	3.458	58	0.62*	1.52*	0.78
Kaingaroa (NZ)	1.190	5.230	4.759	53.	0.74**	1.72*	1.60
Nelson (NZ)	0.938	3.614	4.552	56	0.82***	1.33	0.60

NS, difference not significant at $p>0.05$; * difference significant at $p<0.05$; ** difference highly significant at $p<0.01$; *** difference very highly significant at $p<0.001$.

the variance parameters and heritabilities are summarised in Table 6.10 and estimates of parameters for spiral grain at ring 5 on both sites, population by population, are given in Table 6.11.

Burdon and Low (1981) made the general comment that comparisons of parameter estimates among groups of populations are probably based on too small a sample to come to any definite conclusions. In particular, there is the potential complication that the additive component of genetic variation within the New Zealand populations could have been inflated by the appreciable differences between what are believed to be their two progenitor populations, Año Nuevo and Monterey. There are, in fact, a number of anomalies in these results, but little can be added at this point that would be anything more than speculative.

On the face of it, important genetic gains could be made in reducing spiral grain, considering the very large coefficients of variation and the high apparent heritabilities. It must be remembered, however, that the information is based on only two points in the stem. Admittedly, ring 2 is frequently the point of maximum spirality and ring 5 provides some indication of the persistence or decline of large grain angles in corewood, but further information about the total pattern of spirality is desirable. Burdon also pointed out that spirality does not have a lower numerical bound of zero (the conventional notation employs sign reversal for a change from LH to RH grain angles), so that the coefficients of variation should be treated with some reserve. In particular, the amount of genotype-environment interaction at the levels both of populations and of families, led Burdon to suggest that a cautious approach would be appropriate, since any extension of this trend on sites of greater diversity would vitiate the remarkable within-site heritability.

It remains to describe three sets of experiments that have used radioactive tracers to determine grain angles in the outer wood of standing trees, that is, experiments during which the trees being studied were left intact, with neither wood samples nor bark being removed.

Arbez et al. (1978) studied inheritance of spiral grain in *Pinus pinaster* and *P. nigra* var. *laricio* by examining mother trees in the forest, their open-pollinated progenies aged 10–14 years, and grafts from the mother trees. For both species grain direction was always LH in both grafted and seedling progenies, but grain angles in the grafts were higher. Heritability was estimated in three ways. Broad-sense heritability was estimated from analysis of variance in ramets, narrow-sense heritability from sib analysis on open-pollinated progenies, and the coefficient of genotypic prediction from regressing progeny means on means of the grafts of parental clones. (Coefficient of genotypic prediction is an estimate adjusted for differences in the expression of variation in spirality between the seedling and graft material.)

The respective values for these three estimates for *Pinus pinaster* were 0.43, 0.16, and 0.34, and for *P. nigra* 0.61, 0.58, and 0.38. Although these results could be considered to be encouraging, Arbez et al. (1978) concluded that more information was required concerning variation of spiral angles along the stem, and changing patterns of spiral grain within trees, before spiral grain should be included among selection criteria for tree breeding.

Radioactive tracer techniques were used by Birot et al. (1980) to examine spiral grain in three hardwoods, *Fagus sylvatica, Quercus petraea,* and *Eucalyptus dalrympleana.* Grain angles varied considerably with height up the stem in all three species, most notably in *Eucalyptus dalrympleana*, but significant correlations were obtained between angles at the various heights examined. In this study large grain angles were seldom encountered in *Fagus sylvatica* and *Quercus petraea*. However, grain angles were larger in *Eucalyptus dalrympleana*, and these were believed to show a low, but significant, degree of genetic control.

The third group of studies (Teissier du Cros et al. 1980) also concerned *Fagus sylvatica*. Forty 100-year-old trees were used to study phenotypic variability between trees of variation in grain angle up the stem, and also to investigate the use of bark characteristics to assess grain angle. Grain angles were measured by radiography at 1 m in all trees and also at 5 and 9 m in 20 of them. Another study used an open-pollinated progeny trial containing 30-year-old trees. This experiment contained open-pollinated progenies of eight parent trees that had been selected because of their spiral grain, and eight progenies of parent trees that had been selected as being early- or late-flushing. Since the latter group had not been selected for spirality they were used as "controls". Each progeny was represented by 10 or 12 trees.

Trees for which spiral grain was measured were classified according to utilisation criteria: less than 6° grain angle for veneer, and less than 10° for sawn timber. Of the 100-year-old trees examined, 32 had grain angle at 1 m exceeding 5°, but in only one did the grain angle exceed 10°. At 1 m, 24 trees had RH grain angles, 6 had LH grain angles, and 10 had straight grain. Irrespective of orientation, mean values shifted towards right-handed spirality from the base of the stem towards the top. The close correlation between grain angles at different heights in the stems is demonstrated in Table 6.12. Incidentally, bark characteristics proved unreliable as indicators of spiral grain since five of the eight trees with grain angles greater than 6° had not been detected before radiography, and these included the tree with the largest grain angles (13°–14°).

In the 30-year-old trees of the progeny trial, no bark deviations were noticeable. Family mean values for grain angle measured by radiography were less than 2.5°, but grain angles as high as 7° were measured in individual trees. Six of the 188 trees measured had grain angles exceeding 6°. Grain direction was LH in 37

Table 6.12. Coefficients of correlation between grain angles at different heights on the outside of stems of *Fagus sylvatica*. (Teissier du Cros et al. 1980)

Height in tree	1 m	5 m
5 m	0.84***	
9 m	0.90***	0.93***

*** Difference very highly significant at $p<0.001$.

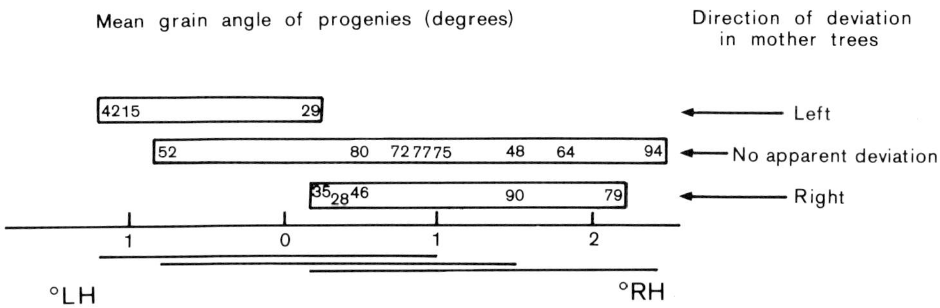

Fig. 6.6. Mean grain angles in progeny trials of *Fagus sylvatica*. Mother trees had been selected as having LH or RH spiral grain, and the "controls" as regards grain angle had been selected purely on the basis of their time of flushing. (Teissier du Cros et al. 1980)

trees, RH in 70 trees, and 81 trees had straight grain. Narrow-sense heritability was estimated to be 0.66, and it was concluded that spiral grain in *Fagus sylvatica* is under strong additive genetic control.

The mean grain angle for each family, grouped according to the observed grain orientation in the mother trees, is illustrated in Fig. 6.6. Apart from family 29, in which grain angles were very small, there was good agreement between grain direction in the mother trees that had been selected for grain spirality, when this was compared with grain direction in their progeny, and the difference between the two groups was statistically significant ($p<0.05$). On the other hand, the mother trees used as controls (selected for flushing characteristics) had progeny with grain angles almost equally widely distributed. Teissier du Cros et al. (1980) concluded that spiral grain had been present in these mother trees but, because bark characteristics are unreliable indicators of spiral grain, it had not been recognised at the time the original trials were established.

6.5 Modification of Genotypes

Under this general heading will be included new combinations of genotypes as in the production of hybrids, the induction of polyploidy, and the genetic consequences of ionising radiation treatments.

Onodera et al. (1977) reported on *Larix* hybrids in Japan. These had originally been produced to study vole resistance of the progeny since the indigenous *Larix*

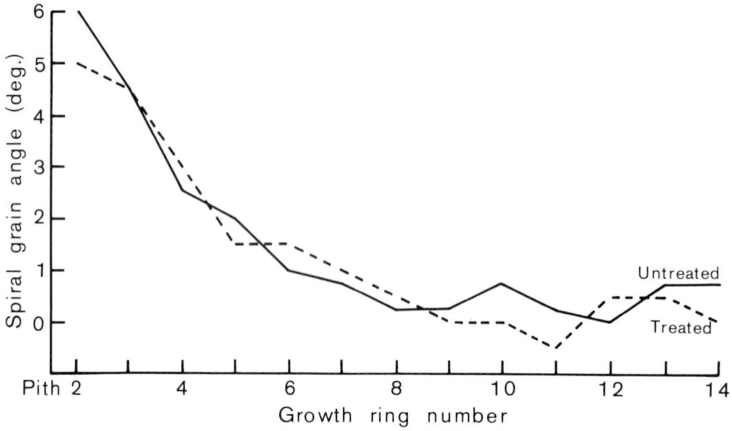

Fig. 6.7. Variation of spiral grain through successive annual growth layers from the pith for colchicine-treated (tetraploid) trees and untreated trees of *Pinus radiata*. (Nicholls and Brown 1972)

kaempferi is particularly prone to vole damage. The species involved were *L. kaempferi* (A), *L. gmelini* (B), *L. olgensis* var. *koreana* (C), and the hybrids *L. gmelini*×*L. kaempferi* (D), and *L. olgensis*×*L. kaempferi* (E). Measurements of spiral grain showed that B>E = C>D>A. Since spiral grain is regarded as being a major defect in *L. kaempferi*, giving rise to serious problems of distortion and twisting in drying sawn timber, there would have to be serious misgivings about using these hybrids (vole-resistant or not!). On the other hand, Smith and Smart (1972) were able to report less spirality compared with the parent species in a hybrid of *Pinus elliottii*×*P. caribaea* var. *hondurensis* grown in Queensland, Australia. The Fl hybrid grew faster than the parents, and had the additional advantages of intermediate wood density and more uniform texture.

Nicholls and Brown (1972) assessed the wood properties of 15 polyploids of *Pinus radiata* aged 18 years. Seedlings had been treated with colchicine and subsequently all the trees examined had been shown to produce pollen indicative of tetraploidy. However, Nicholls and Brown stated that this form of treatment only affects the stem, and the root systems would almost certainly have remained diploid. Among the wood properties studied, wood density was significantly lower in the polyploids than in untreated trees growing in the same area, and radial growth rate was lower in the polyploids over the first 4 years but greater thereafter. There were, however, no differences in spiral grain angles between the treated and untreated trees (Fig. 6.7).

Finally, Haskins and Moore (1933) described twisted growth in two citrus plants (species not stated) after the seeds had been subjected to 24000 R of X-irradiation. Although a number of seeds were irradiated only two plants showed the characteristic abnormalities. Both plants had RH spirality to such a degree that the stem was bent from the vertical, and the leaves were pressed against the stem in early life. However, after 6 months these tendencies ceased, and later growth was normal. Haskins and Moore suggested that this early physiological abnormality may have been the consequence of abnormal mitosis induced by the

X-rays, but it obviously bears little resemblance to the persistence of normal spiral grain.

6.6 Heritability of Wavy and Interlocked Grain

Considering the great value of decorative timbers with wavy and interlocked grain, research into the heritability of these features is disappointingly sparse. In contrast, far more effort has been put into studying the heritability of the various forms of curly grain in the genus *Betula* as described in Section 1.4.3, incidentally, with considerable success towards propagating what are generally regarded as "cambial abnormalities" (Johnsson 1950, Heikinheimo 1951, Persson 1954).

One of the difficulties with all such experiments is that they are necessarily long-term, and both written records and growing trees require conscientious maintenance over many years to obtain useful results. Unfortunately many of the experiments have been initiated by enthusiastic amateurs with the result that final statements all too often have to include some comment about the reliability of the record. For example, Paul (1955) reported that a sapling "thought to have originated from *Juglans nigra* with wavy grain" did not perpetuate this feature.

Considerable interest has been shown in India regarding the possibility of increasing the (normally low) frequency of wavy grain in *Pterocarpus santalinus* because of its value for musical instruments. Kedharnath (1963) described an experiment in which progenies were being raised from both straight-grained and wavy-grained trees. Kedharnath and Rawat (1976) found that fibre lumen diameter was significantly greater in trees with wavy grain, a feature which they suggested could provide a means of screening saplings for future production of wavy grain. Kedharnath et al. (1976) recorded early growth in grafts of *Pterocarpus* derived from ortets with wavy grain but provided no description of the timber. Unfortunately no further results are available from any of these experiments.

One relatively long-term study was reported by Barr (1983) concerning a stand of *Eucalyptus saligna* in New Zealand which had been known to him for 40 years. Natural regeneration derived from three old trees of unknown origin was notable for good stem form combined with excellent timber properties. In particular, two of the original trees had noticeably wavy bark, and second generation trees surrounding these were said to contain "an unusual preponderance" of wavy grain. Seed from a highly figured tree cut in 1974 has been widely planted and recorded to study inheritance of this feature, but no results are available as yet.

The only other reliable report on inheritance of wavy grain is that of Walters (1951) who recorded that wavy grain in *Juglans nigra* has been propagated by grafting.

With reference to interlocked grain, Kadambi (1951) was convinced that this feature is strongly inherited in Indian timbers, and maintained that progeny of trees with strongly interlocked grain could be recognised as maintaining this feature even at the sapling stage. In the United States Paul (1955) noted that clonal material of *Populus canescens* derived from a parent with interlocked grain contained interlocked grain at an early stage of development.

There is obviously need for much more work in this whole area of propagation of decorative timbers.

6.7 Strategies for Tree Improvement

The results reported by Champion (1924–1930) met with an immediate positive response (e.g., Schreiber 1926), and were to influence attitudes in many countries from then on. Reduction of spiral grain through tree selection became one of the objectives of tree-breeding programmes around the world (Büsgen and Münch 1929, Koehler 1939, Dadswell and Nicholls 1959, Phillips 1963, Brazier 1967, Paterson 1967, de Villiers 1973, 1974). Although it is easy, with the benefit of hindsight, to be critical of some of the criteria used to select or reject otherwise superior trees on account of their spiral grain, the general attitude is reflected in the words of Koehler (1949), "To be on the safe side, seed for forest planting should not be collected from trees that have spiral grain. Young trees with spiral grain should be removed from a forest as soon as convenient after they are discovered". However, as Krempl (1965a, 1970) pointed out, discovery may be long delayed for species like *Picea abies* in which the most severe spiral grain develops relatively late in the life of some trees.

Following the experience with *Pinus roxburghii* in India, foresters also became conscious of the dangers of negative selection, in which straight-grained trees were preferred by loggers, leaving the worst of the trees with spiral grain as the seed source for the next generation (Knorr 1932, Mayer-Wegelin 1956).

The difficulty in incorporating wood properties in a tree improvement programme is that, although many properties appear to be *capable* of improvement (Schreiner 1958, Zobel 1961, 1964), they nearly always have to be considered in conjunction with, and are often required to defer to, selection for tree vigour, stem form, and sometimes disease resistance (Wilcox 1978). There may also be problems if spiral grain shows negative genetic correlations with desirable properties such as diameter and height growth, as Allen (1978) noted in *Pinus elliottii*. As a result, some programmes merely try to ensure that wood properties do not deteriorate as a result of improvements in other directions (Smith 1970). Only when trees are to be selected for the decorative value of their timber, are features such as wavy and interlocked grain likely to be given top priority.

As with other traits, spiral grain can be selected by screening provenances, by phenotypic selection of plus trees in wild stands or plantations, by progeny and clonal testing, and in advanced generation selection. There may be differences of opinion about where maximum selection effort can best be applied, but this probably varies with the circumstances. Hughes (1973) argued that it was often impracticable to attempt selection of plus trees in the natural range of a species, and so the policy should be to avoid stands where spiral grain is prominent, and to eliminate trees that develop signs of spirality after seed stands have been established. Wilcox (1978), who was thinking mainly in terms of selection within plantations, took a different view. He maintained that where a feature has high heritability, together with a worthwhile phenotypic variance, but involves a comparatively high cost of assessment (which therefore restricts the selection effort that can be made), it is logical to make maximum use of phenotypic selection before committing trees to progeny testing.

The processes for tree selection have been progressively refined and several attempts have been made to formalise the criteria for plus-tree selection, and for provenance and progeny testing and certification (Champion 1931, Stern 1969, Pederick 1970, 1973). However, the decision to include spiral grain, or any other wood property, in a selection programme is often difficult to justify because improvements suffer from lack of "visibility", i.e., they do not stand out in a forest situation like tree vigour or improvements to stem morphology. Wilcox (1978) suggested that wood quality improvements can best be sold to foresters if they can lift a species over a quality threshold that is perceived to impede full utilisation. Whatever decision is made, it is essential that it should be guided by thorough knowledge of variation patterns (to achieve valid assessment), and good understanding of the economic gains that will arise during utilisation.

6.8 Conclusions

1. Spiral grain has been shown to be under a considerable degree of genetic control in many species. However, from observations of coppice shoots under forest conditions, there appear to be some species in which genetic control of spirality is negligible.
2. Distinction must be made between (a) broad-sense heritability (i.e., in vegetatively propagated progenies) which is the ratio of total genetic variance to the phenotypic variance and (b) narrow-sense heritability (i.e., in sexually propagated progeny) which considers the ratio of the additive genetic variance to phenotypic variance.
3. The issue of main interest to foresters is the potential for genetic reduction of spiral grain through tree selection and breeding or vegetative propagation. Genetic gains require not only high heritabilities, but also wide phenotypic variability, and the opportunities (and incentives) to go for effective selection intensity.
4. The existence in many species of marked differences in grain spirality between geographical origins indicates the potential for screening provenances as well as phenotypic selection of plus trees.
5. The main obstacle to including spiral grain in a programme of tree improvement is that priorities for tree improvement in most species will be given to features such as volume production, stem morphology, and disease resistance.
6. It is important to be able to demonstrate that reduced spiral angles will bring economic advantages during utilisation, and to ensure that valid comparisons of spiral grain can be made between phenotypes, such that effective heritabilities are achieved.
7. There has been a serious dearth of research into heritability of wavy and interlocked grain. Considering the high value of decorative timbers with these features, which would justify giving them top priority in a tree improvement programme, this must rate as a priority area for future research.

Chapter 7

Physiological Aspects of Changing Grain Angles

7.1 Requirements

The development of a tree from embryo to maturity is governed by a series of self-regulating equilibria, modified by the tree's internal and external environments. Yet each species and each individual has its own specific developmental pattern, its own norm to which it tends to conform. Physiology is concerned with the means by which these complex processes are regulated, and no apology is necessary for referring only to "aspects" of physiology in the title of this chapter, because no single definitive system controlling grain orientation in trees can be described. At the moment a few of the pieces of the jigsaw puzzle can probably be recognised, but the nature of the total pattern can be perceived only dimly and in parts.

The immensity of the problem is immediately apparent when one reviews the wide-ranging phenomena to be explained, all of which (as defined in Chap. 1) share the features of longitudinal waveforms in wood, dependent on gradual and regular re-alignment of longitudinal elements across successive growth increments, viz:

1. Spiral grain, which is the normal growth pattern in many species (Northcott 1957), and which usually reverses its orientation (LH or RH) only once during the life of the tree.
2. Wave patterns along the stem, having a relatively uniform period (Hejnowicz 1975), that produce frequent reversals of grain direction across a radius. These cover the range from long-wave patterns, recognisable as interlocked grain, that usually affect the entire bole, to short-wave patterns, recognisable as wavy grain, that usually affect only a part (sometimes a very small part) of the bole.
3. Combinations of these various waveforms, as in wavy spiral grain (Pyszyński 1972), or short waves superimposed on long waves (Zagórska-Marek and Hejnowicz 1980), or waves of different length but similar period interacting to produce moiré and chequered patterns (Hejnowicz and Romberger 1979).
4. Other complications presented by features such as the persistent bifurcation of waves (Zagórska-Marek 1985), and the periodic inactivity of cambial domains producing waveforms (Włoch 1985), as described in Section 5.4.2.

Nor can those examples be ignored in which (1) straight grain is the normal condition, or (2) grain deviations, once initiated, are not corrected but seem to become exaggerated by later growth (Priestley 1945).

That the various processes of grain re-orientation can be attributed to a common grouping of anatomical events suggests some degree of uniformity in the way that most of these grain patterns are fashioned, but gives little indication of the underlying factors regulating these processes. As a starting point it may be helpful to consider those features of stem organisation that determine normal wood structure.

7.2 Polarity

Sinnott (1960) observed that tree growth is organised around an axis along which development takes place, and in relation to which lateral organs are formed. The two poles of this longitudinal axis are unlike, and the polar axis is associated with two groups of phenomena. These are the gradients which appear amongst metabolic processes, and the form and structure of lateral organs that arise successively at the growing point.

The polar axis is also an axis of symmetry. Sinnott saw the radial symmetry that is evidenced in the regular distribution of lateral structures around an axis as arising from an inherent spirality in living things. In fact he recognised four components in the general phenomenon of plant organisation: polarity, differential gradients, symmetry, and spirality. In this respect changing grain angles can be thought of as changes in radial symmetry about the polar axis, arising within a system of differential gradients.

It is within this organisational framework that events, leading to changes of grain angle (or re-organisation of radial symmetry), have to be considered. Some of the key issues include the nature of morphogenetic events, how they are initated, how they are maintained, and what determines their spacing and timing. Timing, in particular, poses special problems because cambial waves move so slowly, characteristically by one wavelength in 10 to 15 years (Hejnowicz and Romberger 1979). The rate of domain migration in wavy-grained *Fraxinus*, for example, may be less than 1 mm year^{-1}. The regulatory process must therefore be capable of maintaining the integrity and highly ordered nature of these events for very long periods of time.

7.2.1 Manipulating Polarity

From early times horticulturists have known that the orientation of phloem and xylem elements can be changed or manipulated by surgical and other means. Knight (1807) described experiments in which bark strips lifted from *Juglans regia* and other species could be induced to generate new cells and "that these may be made, by appropriate management, to traverse the new cellular substances in almost any direction". That is to say that when the bark strip was disconnected at the top, but continuity was retained from one side to the other, cellular alignment changed to connect the two points of contact. However, cells "did not readily ascend into it when it was connected with the bark at its lower extremity only".

Later workers further explored these effects. Hartig (1858) removed a strip of bark in a spiral direction and observed that the wood formed above the girdle became aligned in the direction of the spiral. The experiments of Neeff (1914) are described in Section 5.2.6. He found that decapitation of shoots led to re-orientation of xylem elements towards surviving branches. In some places where xylem elements had been axial when the leading shoot had been in place, new elements changed direction by nearly 180° (Fig. 7.1). Teodoresco and Popesco (1915) used horizontal bark bridges across an area of stem from which the rest of the bark and cambium had been removed, to change xylem orientation by nearly 90°.

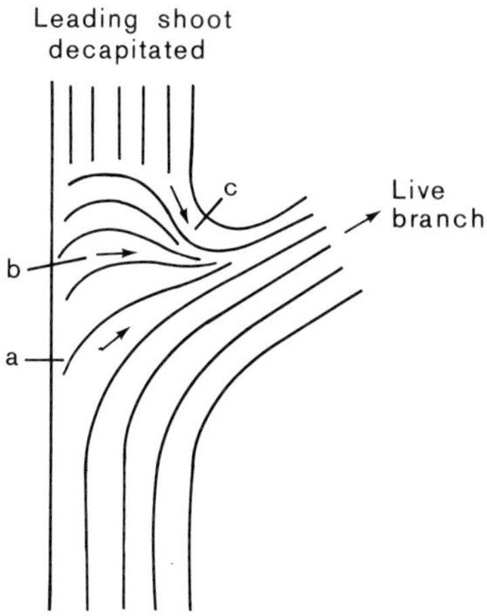

Fig. 7.1 a–c. Re-orientation of longitudinal xylem elements of a decapitated shoot. At (**a**) previously axial elements change direction by 45°, at (**b**) by 90°, and at (**c**) by nearly 180°. (After Neeff 1914)

These and other experiments were later repeated by MacDaniels and Curtis (1930) on trees of *Malus domestica* to observe the processes involved in vascular re-orientation and solute translocation. In one experiment solute translocation was measured after removal of narrow strips of bark extending twice around the stem spirally. Initially, lateral transfer of solutes and metabolites was relatively slow, though it did occur, but then new conducting tissues laid down by the cambium changed their orientation so that the longitudinal axes of the new elements were parallel to the spiral, and localised effects of ringing on solute transfer disappeared. They noted that structural changes providing for more rapid lateral conduction began soon after girdling. The first-formed vessel elements were connected through their radial walls, but the cambium itself later changed its orientation to be parallel with the spiral girdle, and normal connections were then established through the end walls of vessels.

MacDaniels and Curtis believed that the direction of movement of "foods and nutrients" played an important part in the re-orientation of cambium. However, they observed that although cambial cells could be turned through as much as 90° without serious disturbance, it was difficult to turn them through greater angles. Experiments with N-shaped bark bridges, for example, caused serious cambial disturbance, which led them to believe that there were polar factors involved in cambial re-orientation, over and above the influence of the flow of metabolites. Similarly, Collins (1930) used grooves cut through the bark of *Castanea sativa* in an attempt to change xylem orientation by angles up to 100°, but the maximum change recorded was over a distance of 50 mm where grain had "turned almost at a right angle (i.e., 87°) from the perpendicular".

Kirschner et al. (1971) used various modifications of a bark bridge across girdled stems of *Robinia pseudoacacia* (Fig. 7.2) to examine the processes of xylem re-orientation and the factors controlling it. When the bridge was complete (Fig. 7.2 A) newly formed vessels aligned themselves approximately with the components of the bridge. When the bridge was detached from the root (B) fewer vessels were formed, but they had a similar alignment, connecting eventually to

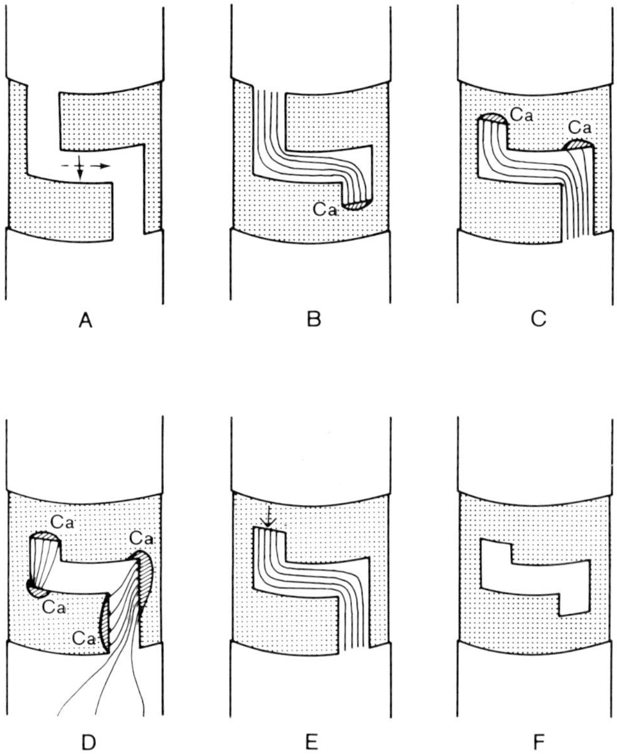

Fig. 7.2 A–F. Xylem re-orientation under a bridge of bark across a girdled area of stem in *Robinia pseudoacacia*. (Kirschner et al. 1971). **A** Bridge complete. The *solid arrow* indicates initial orientation of the xylem, and the *dotted line* that adopted after the bridge was made. **B** Bridge detached from the root. *Ca* indicates callus formation, and *fine lines* indicate new vessels formed. **C, D** Bridge detached from apical part of the shoot. **E** Bridge detached from apical part of the shoot, but 1% auxin in lanolin applied to the cut upper surface. **F** Bridge detached from both root and shoot apex

a zone of callus that had formed where the bridge had been cut. When the bridge was detached from the shoot (C, D) even fewer vessels were laid down and these either connected the lower stem to regions of callus, or made connections between different regions of callus. When the bridge was detached from both root and shoot (F) growth appeared to come to a stop, but microscopic examination showed that the few cells that had been produced lacked any axial orientation. Callus formation in circles at the ends of wood rays, and circular vessels, were two of the features that were observed.

The apparent ability of callus tissues to replace (partially at least) the effect of both root and shoot on xylem re-orientation, raises questions about surgical processes in general, and whether or not some of the changes observed as the result of girdling experiments might more properly have been attributed to wound responses. However, identical results to spiral girdling can be obtained without any bark incisions, and these arise, in nature, from twining vines (Lutz 1943, Harris 1967, 1969). Stem growth is severely restricted and may cease under the constriction of a vine. Vascular development between the encircling coils then rapidly adopts the spiral orientation. Figure 7.3 a illustrates the early stages of this type of development in a young *Pinus radiata* stem bound spirally with heavy electric flex. The initial LH spiral can be seen in the upper part of this stem, and the tendency for grain direction to follow the spiral is most marked in the swollen ridges above the constriction. However, the influence of a live branch on grain

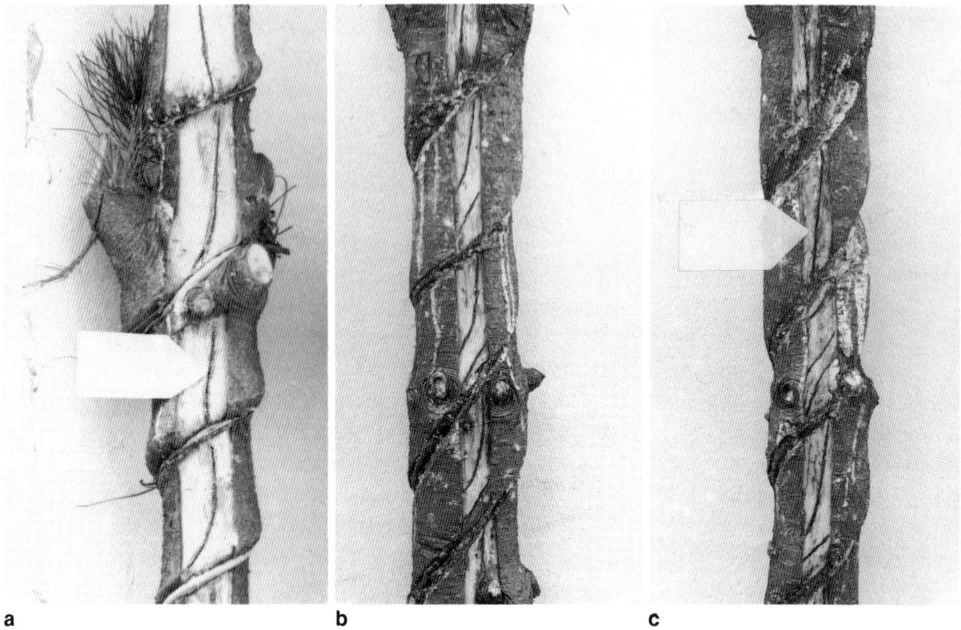

Fig. 7.3a–c. Manipulation of grain angle in *Pinus radiata*. (Harris 1969). **a** Unpruned stem spirally bound with heavy electric flex. *Arrow* indicates region below a live branch. **b** Pruned, spirally girdled stem. **c** Pruned, spirally girdled stem. *Arrow* indicates region below the point of application of IBA (1% in lanolin)

direction below it (arrow) is also clearly indicated. A pruned, spirally girdled stem at a similar stage of xylem re-orientation is shown in Fig. 7.3b.

7.2.2 Patch Grafting

When a patch of bark is carefully removed from a tree during the height of the growing season, separation takes place primarily in the zone of newly formed xylem cells, leaving the cambium intact (Brown and Sax 1962). This enables grafted patches of bark and cambium to be rotated through various angles, or to be transferred between compatible trees.

Brown and Sax (1962) demonstrated the need for pressure to be applied to bark strips separated from the stems of *Populus trichocarpa* and *Pinus strobus* if the cambium were to function normally. In the absence of pressure, the cambial zone (if protected against desiccation) proliferated into an extensive parenchymatous callus, in which new phellogen and cambium eventually differentiated, ultimately producing a new stemlike structure. Brown (1964) later examined the effect of pressure on cambial cultures in vitro, and was able to demonstrate that very low pressures (0.05 atm) were sufficient to bring about distinct differences in patterns of cell division and maturation, whereas cambial explants grown without pressure simply developed an unorganised callus mainly of undifferentiated parenchyma cells.

The first experiments to change polarity by patch grafting were made by Sax and Dickson (1956). They used *Malus* spp. (ornamental crab apples) one variety of which had red wood, and one with white wood. A complete ring of "red" bark and cambium was applied to a "white" tree from which a corresponding ring of bark had been removed. When the graft was correctly orientated, growth beneath it was as vigorous as in the rest of the stem, and red wood was produced. Only at the seam in the ring of bark was a strip of white wood produced by regenerated cambium from the host tree. When the grafted ring of bark was placed upside down, wood formation by the graft was greatly reduced, and the tree was temporarily dwarfed until the normally polarised bridge of white wood at the seam of the graft was able to take over the function of conduction.

The next experiment used two rings of bark grafted in reversed polarity, but with their seams on opposite sides of the stem. The strips of white wood formed at the seams, although not in direct contact, were nevertheless supplied with nutrient sap by some degree of lateral transport. When the two grafted rings were separated by a ring of host tree bark, the ploem and xylem within it were observed to re-orientate in a way that connected the bridges of white wood formed at the seams of the grafts.

Curiously, the inverted bark and the vascular derivatives of its cambium appeared to retain their reversed polarity in these experiments. This result led Thair (1968) to investigate polarity of patch grafts in several species including *Malus* spp. (ornamental crab apples), *Thuja occidentalis, Cornus stolonifera,* and *Sorbus aucuparia.* Square patches of bark and cambium were removed and replaced, or exchanged with compatible pieces from other trees, at orientations of 90° or 180°, together with some patches in normal orientations. Grafts in the three orientations were successful in all species, and the imposed orientation of the cambium was maintained. In some species inverted grafts produced vigorous growth, whereas in others it was retarded, but greatest interest was attached to the patches that were rotated through 90°. The suprising result was that the cambium continued to function normally in the altered orientation, and that secondary xylem and phloem continued to be formed in a small block orientated at right angles to those of the host plant. Where the experiment was allowed to continue, this development was maintained through several growing seasons, and spacing of growth rings in the xylem produced by the grafts corresponded with those of the host plant.

There were a few anatomical abnormalities initially. In some species, for example, vessel elements were shortened, and perforation plates were located in radial walls relative to the axis of the scion. By the second growing season, however, these abnormal developments had largely disappeared and wood formation appeared to be normal. In discussing vascular connections to the patch grafts, Thair and Steeves (1976) stated that the margins showed considerable twisting of elements in a complex pattern. Attempts to interprete these patterns through the use of tangential sections ran into technical difficulties which prevented adequate interpretation, but it was quite evident that nutrient supplies adequate for normal growth must have been transferred through the grafts without changing their orientation.

7.2.3 Implications for Polarity

Three issues have been raised by the various experiments to examine the nature of stem polarity. These are the importance of the physical environment (pressure), the importance of the direction of nutrient flows, and the inherent polarity of cambium that maintains its orientation and integrity in patch grafts.

In his summation of experiments to examine the influence of pressure on wood formation, Brown (1964) expressed the belief that applied pressure is necessary to prevent the loss of cambial initials and obliteration of the phloem (which leads to proliferation of a parenchymatous callus in detached bark) and that this justifies the conclusion that mutual cell forces are essential in maintaining normal differentiation of xylem and phloem. He also believed that pressure has an effect on secondary wall formation. As long as proliferating cells are undergoing cytokinesis and are extending primary walls, no secondary thickening occurs and the cells of the callus remain potentially meristematic. If these cells are brought under external pressure, however, continued cell division and enlargement are inhibited, and deposition of secondary wall commences. Brown concluded that in the search for endogenous mechanisms controlling differentiation at any level, one should not lose sight of the fact that hereditary control of organised development evolved in a physical as well as chemical environment.

Turning now to the experiments designed to modify the direction of flow of metabolites down the stem, their success in changing the orientation of xylem and phloem, with or without drastic surgery, provides overwhelming evidence that polarity is strongly influenced by the direction of flow. At times that rate of response is quite remarkable. For example, Harris (1973) reported that grain angle in *Pinus radiata* had been changed from 4° LH to 38° RH over a period of 9 weeks as a result of spiral girdling.

All of the experiments referred to in Section 7.2.1 were consistently able to modify grain angle (at least up to angles of nearly 90° to the stem axis) by changing the direction of flow of metabolites, which makes the persistence of non-conforming orientation in patch grafts seem quite anomalous. Indeed, Steeves and Sussex (1972) reported on further experiments in patch grafting, using circular patches so that the effects of angles as small as 30° between the graft and host tissues could be studied. Even at this small angle, the cambium retained the orientation given to it and laid down secondary vascular elements in this orientation "with almost incredible accuracy".

The obvious question to be answered is "How does a patch graft retain its orientation *against* the general direction of metabolite flow within the host plant, when it derives the metabolites essential for its survival and continued cell production *from* the host plant?" In fact, how is it that the local area of anomalous polarity is not "rationalised" and smoothed out in conformity with the generalised polarity of the host? The answer would seem to lie in two further observations by Steeves and Sussex (1972) with regard to the anatomical connections established by the graft, and responses in the surrounding host tissues.

Both in the square and in the circular grafts only the original top and bottom of the patches had established a true continuity in the cambium, with resulting continuity in the derived xylem and phloem. The other edges of the graft had united with the stock only by the formation of callus. In fact it was possible to tell in which direction grafts had been rotated, because the original upper ends of grafted cambial initials had made contact with the bases of stock initials, and bases of graft initials with the tops of stock initials. Steeves and Sussex emphasised this as being strong evidence of an inherent polarity in the cells of the cambium.

On the other hand, the cambium of the host or stock showed considerable reorientation in the vicinity of the graft, with the result that "the graft was largely bypassed in the transport system" (Steeves and Sussex 1972). They went on to suggest that the grafted patches were incorporated into the transport system in a special way, such that they were supplied with materials and were not subsequently subjected to the kind of stress which, in other experiments, has resulted in reorientation.

This seems to be a logical explanation, and the only argument might be with the suggestion that patch grafts are incorporated "in a special way". In many respects a grafted patch becomes analogous with lateral organs such as leaves, branches, and flowers. Despite their connection to the vascular supply, the cambium of the stem also becomes orientated around them so that they are largely bypassed by the primary supply lines of the stem, and this is irrespective of whether the lateral organ is a nett exporter or importer of metabolites.

Phillips et al. (1981) demonstrated that the equation of fluid mechanics describing the two-dimensional laminar flow around an object embedded in a fluid provides a very accurate representation of grain angles around knots, so that metabolite flow around the "obstruction" of a graft is sufficient to explain the grain diversion. The likely sequence of events during the establishment of patch grafts is also fairly clear. It is some time, probably more than one growing season, before a patch is able to assume all the characteristics of normal growth, whereas grain deviations in the host stem caused by the initial surgery can be achieved within weeks. Once the by-pass configuration has been established, the organisational pattern of the host stem is well adapted to proceed with the important business of metabolite conduction and the "intrusion" is left to look after itself.

If this is the correct interpretation, there still remain three important aspects of stem polarity. These are (1) the physical environment, particularly mutual cell pressures, (2) stem gradients, especially those represented by the direction of flow of metabolites, and (3) an inherent polarity in the cells of the cambium itself.

The relatively neutral term "metabolite" has been used in this discussion, rather than referring to "nutrient" flows, because many results have pointed to auxin as playing a leading role in determining cambial, and hence vascular, orientation, and this will be considered in the next section.

7.3 The Role of Auxin

Certain experiments to manipulate stem polarity by changing the direction of flow of metabolites also investigated the use of an exogenous source of auxin to simulate the inputs from stem and branches. For example, Harris (1969) had observed that the presence of a live branch within an area of spirally constrained stem (Fig. 7.3 b) could maintain the axial grain direction below it, whereas grain angles elsewhere were conforming with the imposed spiral direction. He therefore applied auxin (in this case, indole-3-butyric acid, IBA) to a section of the spiral girdle in a pruned stem (Fig. 7.3 c) and observed that the axial grain direction was maintained below the point of application, as it had been below the branch, presumably as a response to basipetal flow of the auxin. Similar results were reported by Zagórska-Marek and Little (1986) using *Abies balsamea.*

In their experiments with bark bridges across girdled areas of *Robinia pseudoacacia* stems Kirschner et al. (1971) applied auxin (indol-3-acetic acid, IAA) to the upper part of a bridge that had been detached from the shoot (Fig. 7.2E). Many new vessels were formed after this treatment, in the same orientation as when the bridge was connected to the shoot.

There is, therefore, evidence that auxin is an important component of the metabolites whose direction of flow influences grain direction. Indeed Zajączkowski et al. (1984) went further than this in stating that "among known plant hormones, only auxin exhibits basipetally polar transport. Auxin transport has a striking polarity which mirrors the structural and functional axial polarity of the plant body. Thus, even though other phytohormones are known to cause morphogenic effects, a prominent (and probably dominant) role in the spatial and temporal control of plant morphogenesis must be ascribed to auxin, and especially to the still obscure processes comprising the mechanisms of its basipetal transport".

Major emphasis in the following sections will therefore be placed on auxin as the substance primarily responsible for those changes in radial symmetry about the polar axis that give rise to spiral grain. At the same time, one must be acutely aware of the complex interactions that are possible within the wide range of plant growth regulators influencing vascular development (see, for example, the excellent recent review by Savidge 1985). That it is currently impossible to produce a more comprehensive picture of the interaction between auxin and other growth regulators in changing grain angles is an inevitable consequence of our inability to explain the mode of action of any plant growth substance in any system (Savidge and Wareing 1981).

7.3.1 Auxin and Cambial Activity

The general effects of auxin on cambial development have been summarised by Philipson et al. (1971) and by Savidge and Wareing (1981). Cambial activity in trees corresponds both in space and in time with the distribution of auxin throughout the tree. Even so, there may be differences between different classes of trees. For example, the duration of xylem and phloem formation in relation to auxin levels differs between ring-porous and diffuse-porous trees, and there may be differences between gymnosperms and angiosperms in the ways that auxin controls radial expansion of cambial derivatives.

The classical experiments with seedling hypocotyls suggested that polar transportation of auxin was the result, rather than a cause, of longitudinal polarity in plants (Goldsmith 1969). However, recent research in herbaceous plants has indicated that auxin movement through tissues determines both the occurrence and axis of vascular differentation (e.g., Sachs 1981, Gersani 1985). Similarly, Wareing et al. (1964) using *Fraxinus excelsior* and Egierszdorff and Tomaszewski (1973) using *Pinus sylvestris* have shown that auxin is a controlling factor in xylem differentiation in woody plants. Wareing et al. (1964) studied the different functions of IAA and giberellic acid (GA) in activating the cambium, using both coniferous and dicotyledonous trees. GA and IAA were both capable of promoting cambial activity, but only IAA was able to induce the differentiation of xylem derivatives.

Wodzicki and Wodzicki (1973) showed, for *Pinus sylvestris*, that IAA moves basipetally through tissues external to the xylem, and that, without basipetal supply of substances produced by the crown, no mechanism operated by the roots or the remaining stem tissues near the base of the tree can ensure the required level of auxin in the cambial region, or activate and maintain cambial division. Nix and Wodzicki (1974) showed that $IAA^{14}C$ applied to *Pinus echinata* became radially distributed along a steep gradient from a maximum in the cambial zone into the differentiating xylem. Incidentally, this distribution varied slightly during the growing season in relation to the seasonal transition to latewood formation in intact trees. All the indications therefore point to IAA moving basipetally in the region about the cambial zone as the primary determinant of stem polarity.

A curious sidelight on the influence of endogenous auxin production was reported by Harris (1969). In one experiment, originally designed to reduce auxin production in young (5-year-old) *Pinus radiata*, all stem apices (growing shoot tips) were removed at 2-week intervals throughout a growing season. The experiment misfired because the most obvious result was a massive production of epicormic shoots, removal of which resulted in partial defoliation as successive buds were nipped out. It was therefore postulated that the relative amount of auxin to total photosynthate had actually been increased as a result of this treatment. All three stems so treated reverted to straight grain in one growing season, although they would normally have had strong LH spirality at this age.

7.3.2 Wave Patterns in Auxin Transport

Control of cell alignment by the direction of flow of auxin is, however, still a long way removed from explaining the wide-ranging phenomena related to spiral, wavy, and interlocked grain as set out in Section 7.1. Hejnowicz and Romberger (1979) speculated that morphogenetic waves in the cambium actually only reflect interference patterns of more nearly fundamental and much more rapidly moving waves that are several orders removed from the events that they were able to observe directly.

There are obviously annual cycles of events potentially related to stem polarity, including the annual cycle of auxin production, annual changes in bark pressure, or the geometrical consequences of circumferential expansion. However, these hardly add up to the "much more rapidly moving waves" needed to create the interference patterns postulated by Hejnowicz and Romberger. It is relatively recent-

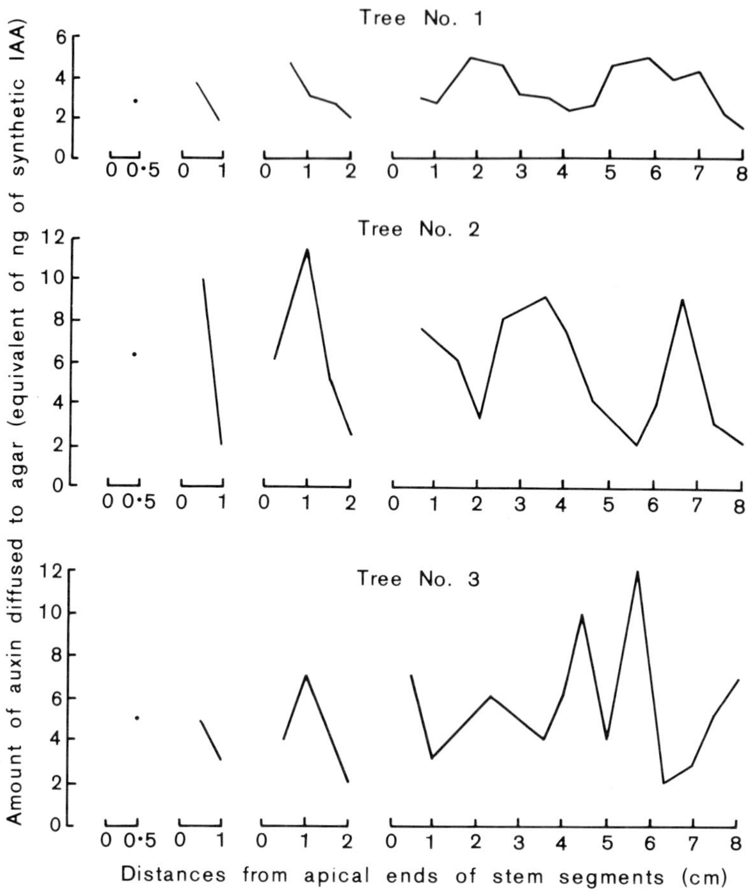

Fig. 7.4. Natural auxin efflux to agar from the basal ends of 5-mm-long sections of pine stem obtained by successive cutting of 5-, 10-, 20-, and 80-mm-long original stem pieces, 10 min after applying agar plates to the basal and apical ends. Results of three experiments performed on different trees: 27 May (*1*), 3 June (*2*), and 10 June (*3*). (Zajączkowski and Wodzicki 1978a)

ly that such wave patterns have been discovered with respect to the amounts of diffusible auxin passing down tree stems.

Zajączkowski and Wodzicki (1978a) measured auxin efflux to agar from the basal ends of stem sections 5 mm long. These in turn had been cut successively from blocks of various lengths, taken from 75-year-old *Pinus sylvestris* trees growing in a forest stand. The results (Fig. 7.4) showed that the amounts of auxin collected formed a wavelike pattern down the stem. The wavelengths were about 20–40 mm long, which is 6–12 times the length of the fusiform cells in the cambial region (ca. 3.4 mm).

Considering that auxin transport occurs in more than just the layer of cambial initials, the waveform of auxin is, to this extent, three-dimensional (Zajączkowski and Wodzicki 1978b). It it also apparent that tangentially neighbouring cells of the conducting zone are at the same phase of oscillation even though their ends

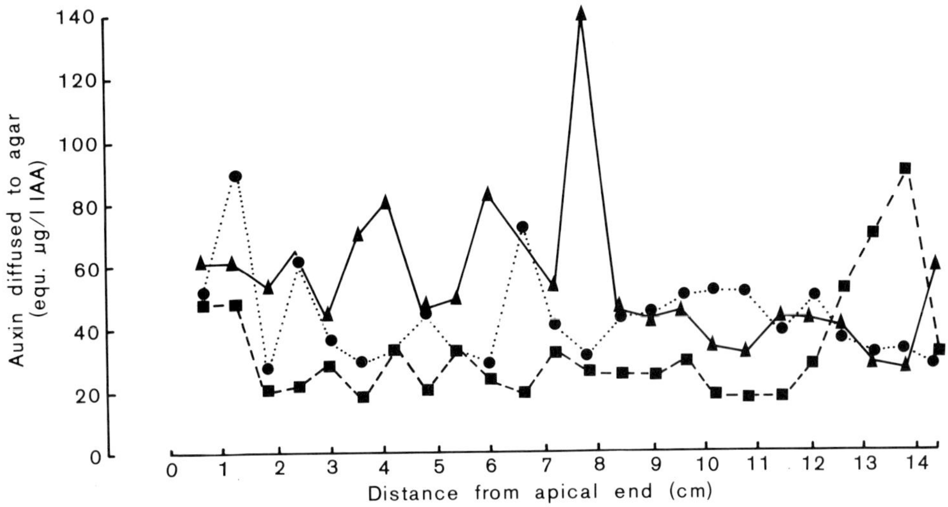

Fig. 7.5. Effect of apical application of IAA (*solid line*), IAA + ABA (*dashed line*), and no additional hormones (*dotted line*) upon the wave of basipetal auxin efflux from consecutive 6-mm sections cut from 144-mm-long blocks of *Pinus sylvestris* stem. (Wodzicki et al. 1979)

do not lie in horizontal rows in non-storeyed cambium. This suggests integration of cells into a supra-cellular oscillatory system.

Wodzicki et al. (1979) further investigated the wavelike pattern of auxin efflux in *Pinus sylvestris* and found that the amplitude of the auxin wave is increased by application of IAA to the longer stem sections before subdivision into 5-mm test pieces. This effect was particularly marked at the time of initiation of cambial activity in the spring. The waveform of auxin efflux was suppressed when IAA and abscisic acid (ABA) were applied simultaneously to the apical end of the longer stem sections before subdivision (Fig. 7.5).

Wozicki and Wodzicki (1981) pursued this matter further in other species and discovered the same wavelike pattern of auxin efflux from dissected stem sections of *Larix decidua*, *Picea abies*, and *Acer pseudoplatanus* as did Zakreski (1983) from stem sections of *Quercus robur*, and Wodzicki et al. (1987) from sections of *Pinus contorta*. Incidentally, Wodzicki et al. (1987) used gas chromatography-mass spectrometry to demonstrate beyond doubt that the wavelike pattern of auxin, diffusing from both *Pinus sylvestris* and *P. contorta*, consists primarily of IAA.

Wodzicki and Wodzicki (1981) went on to investigate the effects of several other substances that might modulate the auxin wave pattern in *Pinus sylvestris*, and these proved to be highly effective in physiological concentrations. Zeatin (ZEA) or gibberellic acid (GA_3) blocked the repression of the auxin waveform by ABA. 2,3,5-Triiodobenzoic acid (TIBA) behaved in the same way as ABA in suppressing the wave pattern of auxin efflux. They also found that ABA applied to the basal end of the stem section reduced the auxin wave amplification caused by simultaneous application of IAA to the apical end. Thus, modulation of the auxin wave can be propagated acropetally, that is, against the main direction of auxin molecular transport.

The results of applying growth substances to stem sections also suggested that modulation and amplification of the auxin oscillatory system could not be attributed simply to translocation of the growth substances involved, because the effects travelled much faster than their known translocation velocities. This supposition was confirmed experimentally for amplification of auxin waves by apical application of ^{14}C-IAA to pine blocks, after which bioassays for hormone activity were combined with determinations of radioactivity (unpublished experiments of Wodzicki and Brown, reported by Zajączkowski et al. 1984). This was interpreted as indicating that the cells participating in the polar transport of auxin functioned as an integrated supracellular system of coupled oscillators, and that the apical meristematic centres were comparable to pacemaker regions (Zajączkowski et al. 1984).

7.3.3 Auxin and Morphogenesis

Before the discovery of wave patterns in auxin efflux from stem sections, Hejnowicz (1975) had proposed a model for a morphogenetic map and clock based primarily on his observations of waveforms in cambial morphogenetic phenomena. He suggested that invariance of period was the clue to understanding the nature of morphogenetic waves, and that the only basis for the formation of such waves would be a population of oscillators, each oscillating essentially by itself, but with some phase shift in relation to its neighbours. The period would be determined by the nature of the oscillators. The velocity, and hence wavelength, would be governed by the interaction.

Hejnowicz (1975) then proceeded to consider how genes might provide an organism with both maps and clocks for morphogenesis. In this model the changing phases of oscillations within the morphogenetic map gave topographic information. A rather different concept was later put forward by Zajączkowski and Wodzicki (1978 b) in which positional information was ascribed to vectors of morphogenetic wave propagation, and deflections of these vectors from the axes of polarity of biological units were recognised by the responding cells. A three-dimensional vector field would behave similarly to other physical fields, and where local spatial inhomogeneities occur this type of field could exhibit local turbulences and aberrations capable of influencing growth and differentiation processes. Harris (1973) had also come to the conclusion that vector components provided the most likely explanation for cellular response to auxin flow.

As to the nature of the protoplasmic oscillators, Hejnowicz (1975) believed that they might be of different types, such as macromolecules, multi-molecular elements, or even cells. However, he conceded that a period lasting many hours is likely to be beyond the range of an individual oscillator, and it was clearly impossible that they should relate directly to the multi-year period of cambial morphogenetic waves. On the other hand, the longer periods could arise in response to beats resulting from wave superposition, because the frequency of the beat will be much lower than the component oscillations. Indeed, if the amplitude of the oscillation affects a certain process non-linearly, the beat will result in a second-order oscillation in this process, with the possibility for a second-generation reduction of wave frequency.

Naturally, the discovery of auxin waves and their modulation by other growth regulators was seen to provide support for these hypotheses of morphogenesis. Zajączkowski and Wodzicki (1978b) set out some general concepts as a basis for further studies. Inter alia, they supported the idea that three-dimensional morphogenetic waves originate in individual cells as a result of synchronised oscillatory processes within them. They believed that the regulatory function probably required some minimum cellular mass, and that spatial relationships giving rise to tensile and compressive forces might modify the direction of vectors of morphogenetic waves to induce specific differentiation of cells. Irrespective of the other functions of various growth regulators, they saw their role in morphogenesis as being primarily a result of their effects on auxin wave propagation, because all are known to influence auxin transport. Thus, as noted in Section 7.3, the exceptional position of auxin among all other growth regulators was attributed to its polar transport, and the function of other growth regulators with respect to morphogenesis was seen to result primarily from their influence on this transport.

Zajączkowski and Wodzicki (1978a) discussed the applicability of these concepts to the development of cambial morphogenetic domains. Auxin transport in the stem is broadly parallel to the longitudinal orientation of cells. Auxin waves also develop in this direction and consequently all points that are in the same phase are located on lines perpendicular to the vector of auxin basipetal transport (the vector of wave propagation) and probably also perpendicular to the polarity gradient of the cells. The spatial pattern of these lines can be envisaged as a map of stem polarity within the cambial envelope, which can be considered to be a type of morphogenetic map, four simple models of which are presented in Fig. 7.6.

Model A is based on the total uniformity of action by all possible factors determining the direction of auxin waves. In this situation all cells are parallel to the stem axis.

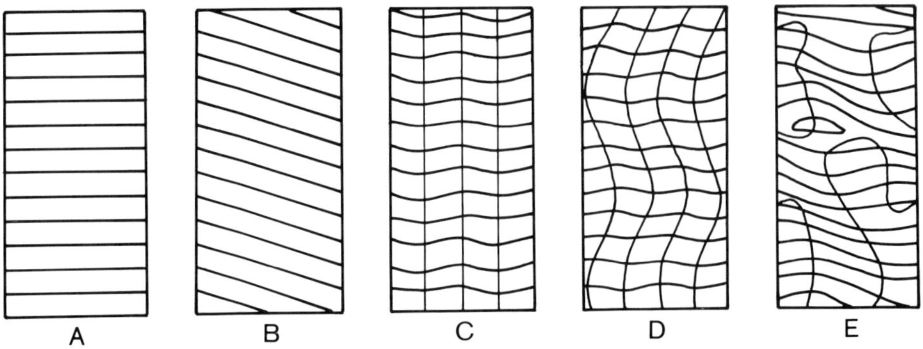

Fig. 7.6. Models of morphogenetic maps based on wave fronts of auxin transport (*thick lines*), and showing domains of right and left inclination of vectors of wave propagation (delimited by *thin lines*). The vectors of wave propagation are always strictly orthogonal to the wave fronts. All models represent tangential views of cambium around the circumference of stem sections. For detailed description see the text. (Zajączkowski and Romberger 1978a)

If one unilaterally modifying agent prevails, the effect is as shown in Model B. In this case the axes of all cells deviate from the stem axis by the same angle, forming a helical pattern of cell orientation in the cambium.

Model C envisages the action of more than one modifying agent. These are of equivalent effectiveness and are distributed equidistantly around the stem circumference. Model D introduces additional equidistant modifying agents along the stem axis. In both of these models inclination of cells to the stem axis is expected to repeat periodically, forming domains of left and right inclination similar to the cambial domains described in Section 5.4.

Finally, Zajączkowski and Wodzicki (1978a) considered all the agents that might modify the orientation of wave fronts. These would include auxin sources, auxin acceptors, processes affecting auxin concentration and transport (e.g., other growth regulators), branching of the stem axis, and so on. All of these are under genetic *and environmental* control, and may be seen as causing complex morphogenetic maps of the type illustrated in E.

The results obtained by Wodzicki et al. (1979) had provided evidence in favour of apical control of the morphogenetic field by influencing a supracellular system of conjugated oscillators. In addition, acropetal modulation of the auxin wave pattern by ABA (Wodzicki and Wodzicki 1981) was seen as further support for the model of morphogenetic fields proposed by Zajączkowski and Wodzicki (1978a) which required modulation of wave parameters in directions other than basipetally. These aspects were also summarised by Wodzicki (1980) and Zajączkowski (1980).

The mode of action of growth substances in producing morphogenetic changes has still to be explained. It is therefore of interest that Das and Dayanand (1983) observed that tyrosine is present in the cambial region of *Pterocarpus santalinus* with wavy grain, but is absent from the cambium of straight-grained trees. Thus the possibility of chemical mediation in the re-orientation of cambial initials may have to be considered.

7.3.4 Response of Cambial Initials to Auxin Vectors

In Section 5.2.6 reference was made to the various means by which fusiform initials appear to change their orientation. It was argued that there were some circumstances in which pseudotransverse division followed by cell elongation, and/or imperfect periclinal division, seemed inadequate to explain all the observed changes in orientation and configuration of fusiform initials. The opinion was expressed that fusiform initials could change their orientation by differential growth between periclinal divisions. The reason for returning to this point now is that some of the observations of this type of re-alignment have particular relevance to the concept of cellular response to auxin vectors.

Paradoxically, sufficiently frequent re-alignments, of magnitude capable of measurement, can be achieved experimentally only under circumstances which must raise questions about their validity under normal conditions of growth. When stems are girdled, or when growth is constrained, in such a way as to induce lateral movement of auxin, the inital cambial response is an increase in the frequency of anticlinical division without any change in the pretreatment ratio of S- to Z-oriented events (Kirschner et al. 1971, Hejnowicz 1972, Harris 1973, Savidge and Farrar 1984, Zagórska-Marek and Little 1986). This response was attributed by Zagórska-Marek and Little (1986) to the supply of auxin being abruptly reduced. They suggested that the eventual change in the ratio of S- to Z-orientated anticlinal division, together with intrusive growth events in favour of the imposed alignment, commences when the new line of auxin flow induced by the treatment actually begins to be "perceived" in time and space. If this interpretation is correct, aspects of fusiform initial re-orientation can still properly be interpreted as responses to auxin vectors.

Harris (1973) measured changes in orientation of both fusiform initials and groupings of ray initials, after spiral girdling of *Pinus radiata*. The results are shown in Table 7.1. After girdling, fusiform initials progressively changed their alignment towards the direction of the RH spiral girdle, by 7° after 3 weeks, by a further 14° after 6 weeks, and by a further 20° after 9 weeks. Measurements were made on tangential sections of the cambial zone cut from samples collected 6 weeks after girdling. Anticlinal division and intrusive growth of initials were estimated to account for only 0.2° mm^{-1} of radial growth, compared with re-orientation amounting to at least 10° mm^{-1}. However, the possible importance of imperfect periclinal divisions was not considered, and no estimate was made of their frequency or significance during this experiment.

The tendency for cell tips to develop a greater angle to the stem axis than the main body of the cell creates a cell with slight sigmoid curvature at the height of the re-orientation process. This was interpreted as a tendency for cell tips to grow differentially and adjust first to changes in the direction of auxin flow, but, because the average angle between cell tips and the main body of the cell never exceeded 5°, it was postulated that some adjustment of the cell axis took place in the period of cell growth after periclinal division. This adjustment would relieve stresses arising from the sigmoid curvature by straightening up between the cell tips. Re-orientation is envisaged, therefore, as arising through differential growth of cell tips in response to change direction of auxin flow, with partial adjustment of the cell axis after each periclinal division. The passive role of ray initials is illustrated by the way they lag behind fusiform initials in re-orientation, and the larger the aggregation of cells in a ray, the greater the lag.

It is apparent that imperfect periclinal divisions might have contributed to these results, but there are other aspects of cell re-alignment, in normal development of spiral, wavy, and interlocked grain, that clearly indicate the capacity for differential growth by the tips of fusiform initials. As mentioned in Section 5.2.6, "creeping" of the ends of fusiform initials was seen to be the major feature of grain re-alignment in the storeyed cambium of *Entandrophragma cylindricum* (Hejnowicz and Zagórska-Marek 1974). However, the clearest indication of differential growth is the intrusion of cell tips between ray initials in the phenomenon of ray splitting. This can occur independently of pseudotransverse division, and cannot be explained in terms of imperfect periclinal division. Ray splitting requires differential growth of considerable force, and if it can develop in this situation it does not seem unreasonable to assume that it can do so elsewhere. Hejnowicz and Zagórska-Marek (1974) described the results of chang-

Table 7.1. Angle to the stem axis of cambial elements of *Pinus radiata* observed at various periods after spiral girdling. (Harris 1973)

Weeks after girdling	Statistic	Fusiform initials				Ray initials		
		Overall (tip to tip)	Upper tip	Lower tip	Centre of cell	Multiseriate	Uniseriate	
							5 cells or more	4 cells or less
0 (control)	Av. angle (°LH)	3	3	4	3	3	3	3
	SD	1.35	2.39	2.84	2.01	2.77	2.44	2.85
	n	50	50	50	50	60	60	60
3	Av. angle (°RH)	4	4	6	2	2	2	4
	SD	3.07	5.52	6.32	2.98	3.67	3.34	4.24
	n	50	50	50	50	65	65	65
6	Av. angle (°RH)	18	20	18	15	10	12	16
	SD	3.99	6.40	6.31	3.95	5.83	5.14	4.09
	N	50	50	50	50	54	79	79
9	Av. angle (°RH)	38	37	37	38	36	37	33
	SD	4.16	5.76	7.98	5.07	6.33	6.63	7.17
	n	50	50	50	50	52	77	83
29	Av. angle (°RH)	50	43	49	50	43	49	51
	SD	2.38	6.16	4.80	2.24	2.93	3.19	5.62
	n	50	50	50	50	70	70	70

ing grain angles in *Tilia parvifolia* in which fusiform initials are storeyed, but rays are taller than one storey. They stated that under these circumstances rays "must be split when the cambium changes the orientation of its cells".

The conclusion that the cambium is very active and plastic, as suggested by Sinnott (1960), is the consequence of accepting that fusiform initials can grow differentially and adjust alignment in response to auxin vector patterns. Harris (1973) made the point that the cambium is not a static tissue, simply cutting off phloem cells to one side and xylem cells to the other: the cambium must also renew *itself* in a successively widening sheath. After each periclinal division the cell that retains the function of cambial initial must grow, mainly by extension of its radial walls, before redividing.

It is now suggested that, in the process of renewing itself, each fusiform initial may also partially re-align itself in conformity with the current information fed to it concerning polarity. Even in a rapidly changing environment (e.g., after a stem has been girdled) the most rapid changes that have been observed in cell reorientation do not exceed (and are usually much less than) 1° for each periclinal division. This would require a maximum adjustment of about 30 μm (one tracheid diameter) at one end of a fusiform initial 2 mm long. Stresses within the cytoplasm, combined with localised stiffening or relaxation along the primary wall, would provide a mechanism for differential growth in keeping with the known activities of growth substances.

7.3.5 Orientation of Pseudotransverse Division

Harris (1973) suggested that the same cytoplasmic stresses that he envisaged as contributing to differential growth in fusiform initials might be responsible for determining the orientation of the new cell wall in pseudotransverse division. Thus, uniformity in orientation of pseudotransverse divisions could be interpreted as being one of the more visible and significant of cellular responses to auxin vectors.

Sinnott and Bloch (1939) emphasised the importance of a band of microtubules that becomes localised in preprophase along the cytoplasmic cortex at the place where the cross wall will fuse with the parent wall during cytokinesis. Lloyd (1984) maintained that this structure, the phragmosome, can lay claim to the title of being the organelle least understood, yet most likely to help our understanding of pattern formation in plants. A perplexing feature of cell plate orientation is the fact that the microtubules of the preprophase band disappear by prophase, long before cell plate formation begins. Hepler (1981) suggested that the band indicates a specialisation of the cell cortex, in particular the plasmalemma, at that site.

Hepler (1981) then went on to consider what factors might regulate and drive rotation of the mitotic spindle apparatus during cell division. Application of various drugs reveals that the motion requires energy, and that it is dependent on the structural integrity on the spindle phragmoplast. Two anti-microfilament drugs specifically inhibit spindle rotation, which provides strong circumstantial evidence in favour of force generation through an actomyosin system.

Whereas these observations stress the importance of cytoplasmic polarisation in cell plate orientation, there is another body of work that tends to emphasise the role of mechanical stress in determining the partitioning of cells in tissues. This raises the question as to whether the preprophase band is the result or the cause of cytoplasmic polarisation.

Lintilhac and Vesecky (1984) suggested that the architectural nature of cell patterns in plants offers strong intuitive evidence for a structural relationship between mechanical forces acting through the tissue and cell wall orientation at cytokinesis. After compressing a callus cultured from *Nicotiana tabacum* for 5 days under compressive stresses of 500 to 800 mg mm^{-2}, they observed coplanar cell division within it, accumulating to a depth of ten ranks of cells. This gave the load-bearing region of the explant an appearance similar to that produced by cambial activity [cf. the in vitro experiments by Brown (1964) described in Sect. 7.2.2].

Lintilhac and Vesecky (1984) saw this result as evidence that extracellular mechanical cues are sensed by cells in tissues, and the new cell plate is orientated so that it is maximally protected from shear. In commenting on this, Lloyd (1984) stated that this work provided further support for the concept that the development of plant form involves epigenetic control over the alignment of cross walls throughout a programme of division and expansion, which imparts to each cell a different positional value that is "read" in terms of strain alignment of the cell plate.

At this point it will be noted that the discussion has moved more towards the topic of plant form, and it may well be that mechanical stress has a different relevance in newly developing tissues (stem apices, for example) compared with its effects on the vascular cambium. Worrall (1980) faced this dilemma in his observations of living stumps in *Pseudotsuga menziesii*. These are kept alive by root grafts to adjacent living trees, and grain direction within them becomes random, changing direction widely over short intervals of time. Worrall was inclined to attribute this development to the lack of the normal stresses associated with the presence of a tree trunk above the stump, but also acknowledged that the hormonal stimuli derived from the host tree would be non-directional. Shoots grafted to the stump to provide a directed flow of auxin unfortunately survived for only 2 years, but did produce some straightening of the grain below them.

The relative contributions of mechanical stress and auxin-induced polarity in determining cell plate orientation in pseudotransverse divisions obviously deserve further study. Evidently both are involved, but in a normal cambial environment the evidence in favour of the orientation of pseudotransverse divisions being largely determined by cytoplasmic polarity arising from auxin vectors seems very strong.

7.4 Other Factors

The promise of recent research should not be permitted to exclude all other concepts, some of which were put forward long before growth substances were even discovered. A few of these can certainly be dismissed quite simply in the light of current knowledge (and sorting the wheat from the chaff is always a useful process), but others may suggest new lines of approach to current problems, or even indicate quite different principles and systems that require further investigation.

7.4.1 Environmental Causes

Environmental causes have been covered rather fully in Chapter 4 and need not be repeated. It may be noted, however, that in considering the agents that would modify the orientation of auxin wave fronts, Zajączkowski and Wodzicki (1978a) acknowledged that all would be under genetic and environmental control (Sect. 7.3.3). Indeed, the influence of auxin on plant growth introduces the possibility of some type of self-regulating equilibrium being responsible for the apparent contradictions concerning the effect of growth rate on grain angles, e.g., abundant auxin induces rapid growth: rapid growth favours wider grain angles (the geometric effect?): strongly basipetal flow of auxin favours straight grain. But the effects attributed to many environmental features, including the effects of wind (torsional stresses), temperature, and eccentric growth (reaction wood), would benefit from further analysis and experimentation in the light of recent results.

7.4.2 Microfibril Angle

Twisting of growing cambial initials as a consequence of microfibril angle in the cell wall (Misra 1939, Preston 1949, Foulger 1966) can be ruled out as a cause of spiral grain, because microfibrils in the primary wall form a network, and do not have a preferred (RH or LH) orientation. Since grain angles are determined by cambial orientation, there can certainly be no connection with microfibril angle in the S_2 layer of the cell wall, which is, an any case, almost invariably RH (Hejnowicz 1964, Meylan and Butterfield 1978, Zagórska-Marek 1984b), and therefore could not account for changes in grain direction. However, the paper by Preston (1949) draws a parallel with twisting during growth of single cells (sporangiophores of *Phycomyces*), and goes on to discuss how a similar twisting growth in a growing fusiform initial might be accommodated within the cambium. This discussion contains a number of features of relevance today, especially with regard to the very small angular changes required between each periclinal division.

7.4.3 Phyllotaxis

The parallel between spiral patterns of leaves or short shoots around young stems, and the development of spiral grain within them has attracted the attention of many writers (e.g., Braun 1854, Bradley 1918, Harris 1969, Krawczyszyn 1973). Modifications of these patterns were shown to influence changing grain angles in *Pinus radiata* (Sect. 4.1.9), and Harris (1969) believed that the needle traces (vas-

Fig. 7.7. Portion of a stem of *Pinus radiata* into which wooden pegs were driven to supplement the original needle traces after the stem needles had been shed. Grain angle in the previous year's growth (here revealed by turnery) is maintained in the area so treated, but decreases in the untreated stem below. (Harris 1969)

cular connections to the short shoots) in *P. radiata* acted, in this respect, as interruptions to auxin flow (Fig. 7.7).

This possibility is difficult to demonstrate mathematically, but was illustrated by recording the passage of steel balls running down an inclined plane into which nails had been inserted in the pattern of the stem parastichies. The distribution of the steel balls after they had traversed this "bagatelle board" was strongly skewed in the direction of the steeper of the primary "spirals". Harris (1969) concluded that although there are obvious limitations for the application of these results to conditions in a tree, they did indicate that the distribution of needle traces around a stem could favour generation of spirality in the flow of auxin, as long as it moves under conditions approaching free flow.

Krawczyszyn (1973) demonstrated that the initiation of cambial domains in *Platanus acerifolia* is associated with the position of the leaf traces in a young stem. However, this pattern disappears by the fourth year's growth.

There is therefore some evidence that spiral grain may be initiated in the young shoot as a consequence of the arrangement of vascular connections to the leaves.

However, this obviously has little bearing on the development of spirality later on, because vascular connections to the leaves are usually occluded very soon after the stem leaves are shed.

7.4.4 Time

The time scale of waveforms in trees still poses major problems. In Section 7.3.3 reduction of wave frequency through beat phenomena was discussed, with the possibility that more than one reductive cycle could be developed. However, a number of features of cambial wave patterns seem to require the introduction of real time into these processes.

Krawczyszyn and Romberger (1975, 1980) provided evidence that biological rhythms are time-dependent in *Nyssa sylvatica* and *Liriodendron styraciflua*. In their studies of pairs of stems growing from a common root system but having different diameters, they were able to show that interlocked grain in these species was related to calendar time rather than to the growth processes controlling radial increment. The observations of Détienne (1979) that interlocked grain in *Daniella* spp. has an annual periodicity would also be difficult to explain as arising coincidentally from interaction of time-independent waves.

So far there is no indication of the nature of the time-dependent phenomenon with which auxin waves might interact, or indeed that auxin waves are involved in the time-dependent process. Continuous growth in a controlled environment might be used to investigate this feature, but it would be difficult to isolate the potential components such as annual cycles in auxin production, bark pressures, and so on. However, the means by which real time might be introduced into the biological rhythms of changing grain angles is possibly one of the most important aspects remaining to be discovered.

7.4.5 Straight Grain and Uncontrolled Grain

Other aspects of spiral grain that have received little attention are those related to the two extreme examples, i.e. species that are almost invariably straight grained, and those in which minor grain deviations acquired in youth are never corrected but appear to become exaggerated by circumferential growth (Priestley 1945).

Bannan (1966a) and Hejnowicz (1968) noted the presence of cambial domains in straight-grained trees. Frequency of morphogenetic events in their cambia tends to be low, and as LH domain is succeeded by RH domain any minor deviations that have arisen in grain direction are soon corrected. This situation deserves further investigation, and comparisons should be made with species in which grain angles change rapidly, particularly with respect to auxin levels in the stem, their mode of generation, and the responses of cambial cells.

The other extreme is, in some respects, even more intriguing. Since Priestley (1945) observed that grain deviations in *Sambucus* and *Syringa* persisted and appeared to increase geometrically with stem diameter, there has been criticism that a labile system like the cambium is unlikely to respond in this way (Newman 1956), but no one has denied that such examples exist, or that their *appearance* is often in keeping with Priestley's hypothesis (cf. Frontispiece). Irrespective of the feasibility of the geometrical explanation put forward by Priestley, the apparent inability of such stems to keep grain deviations within limits that at least give the appearance of efficient conducting systems, is worthy of further study in the light of current knowledge.

7.5 Conclusions

1. This chapter commenced by comparing our current understanding of the physiological processes responsible for spiral grain with a jigsaw puzzle in which only parts of the total pattern can be perceived. It went on to examine some of the more revealing fragments that have been assembled so far, and ended by looking for some of the missing pieces.
2. The premise that changing grain angles can be thought of as changes in the radial symmetry of a stem about its polar axis, arising within a system of differential gradients, has yielded some useful concepts about the nature of these components.
3. There are at least two major components of stem polarity. Tensile and compressive forces exerted by cells within a tissue are an important part of the physical environment. Strongly basipetal flow of auxin dominates the chemical component.
4. It is also probable that fusiform initials in the cambium have an inherent polarity of their own.
5. The very long period (approximately 10 years) of wave patterns in the cambium giving rise to waveforms in wood poses special problems. It has been proposed that the perceived waves only reflect interference patterns of more nearly fundamental and much more rapidly moving waves that may be several orders removed from the events that can be observed directly.
6. The relatively recent discovery of wavelike patterns in auxin efflux from stem sections fits in well with this hypothesis, especially as other growth substances in physiological concentrations have been shown to be capable of modulating auxin wave patterns.
7. Although much work remains to be done, evidence obtained so far favours apical control of the morphogenetic field through its influence on a supracellular system of conjugated oscillators.
8. The mode of operation at the cellular level can be ascribed to a three-dimensional auxin vector field.
9. Fusiform initials may react directly to auxin vectors (changing their orientation by angles less than 1° between periclinal divisions), in addition to the readily recognisable processes of uniformly orientated pseudotransverse divisions followed by intrusive growth.
10. Spiral grain is therefore a topic that offers great promise for future research. Tissues of the cambial zone in large trees can serve as an excellent system in which to study mechanisms of long-distance morphogenetic control. Such tissues have

a high degree of structural and physiological uniformity, lending themselves to studies of polar transport of auxin and its possible involvement in transmitting morphogenetic information.

11. Many older concepts, including those relating to the influence of environment on changing grain angles in trees, require re-assessment in the light of current knowledge.

References

Ahola V K 1952 Mukurapuista. (On gnarly trees). Commun Inst For Fenn 40.18: 1–10

Allen P J 1978 Genotypic and phenotypic correlations of wood and tree characteristics. In: Nikles D G, Burley J, Barnes R D (eds) Progress and problems of genetic improvement of tropical trees. 1: 184–196

Anderson E A, Koehler A, Krone R H 1945 Instruments for rapidly measuring slope of grain in lumber. US For Serv Rep FPL – 1592, 12 pp

Andrew I A, Burley J 1973 Summary report on variation of wood quality of Pinus merkusii Jungh and de Vriese: five trees of Burma provenance grown in Zambia. In: Burley J, Nikles D G (eds) Selection and breeding to improve some tropical conifers 2: 126–132

Arbez M, Baradat P, Birot Y, Azoeuf P, Hoslin R 1978 Variabilité et hérédité de l'angle du fil du bois measuré à l'aide d'un traceur radioactif chez le pin maritime et le pin laricio de Calabre. Can J For Res 8: 280–289

Archer R R 1979 On the distribution of tree growth stresses. 3 The case of inclined gain. Wood Sci Technol 13: 67–78

Bailey L F 1948 Figured wood – a study of methods of production. J For 46: 119–125

Balodis V 1960 Investigation of physical properties and quality of sawn material in plantation-grown hoop pine (Araucaria cunninghamii Ait). Queensland For Serv Res Note 11, 88 pp

Balodis V 1971 Twist in seasoned boards from planation thinnings. Proc IUFRO 15 Would Congr, 15 pp

Balodis V 1972 Influence of grain angle on twist in seasoned boards. Wood Sci 5: 44–50

Balodis V, Sheard R P 1959 Physical properties of timber with relation to growth conditions. Relationship between length of internode and spiral grain in plantation-grown hoop pine. Queensland For Serv Lab Rep QTP 3-2, 16 pp

Banks C H 1953 Spiral-grain and its effect on the strength of South African grown pines. J S Afr For Assoc 23: 45–50

Banks C H 1969 Spiral grain and its effect on the quality of South African timber. For S Afr 10: 27–33

Banks C H 1971 The relationship of log spirality to rotary-cut veneers and plywood and to sliced shooks. CSIR A Afr Special Rep Hout 34, 13 pp

Bannan M W 1950 The frequency of anticlinal divisions in fusiform cambial cells of Chamaecyparis. Am J Bot 37: 511–519

Bannan M W 1951a The reduction of fusiform cambial cells in Chamaecyparis and Thuja. Can J Bot 29: 57–67

Bannan M W 1951b The annual cycle of size changes in the fusiform cambial calls of Chamaecyparis and Thuja. Can J Bot 29: 421–437

Bannan M W 1953 Further observations on the reduction of fusiform cambial cells of Thuja occidentalis L. Can J Bot 31: 63–74

Bannan M W 1954 Ring width, trached size, and ray volume in stem wood of Thuja occidentalis L. Can J Bot 32: 466–479

Bannan M W 1955 The vascular cambium and radial growth in Thuja occidentalis L. Can J Bot 33: 113–138

Bannan M W 1956 Some aspects of the elongation of fusiform cambial cells in Thuja occidentalis L. Can J Bot 34: 175–196

Bannan M W 1957a Girth increase in white cedar stems of irregular form. Can J Bot 35: 425–434

Bannan M W 1957b The relative frequency of the different types of cambial divisions in conifer cambium. Can J Bot 35: 875–884

Bannan M W 1957c The structure and growth of the cambium. Tappi 40: 220–225

Bannan M W 1960a Cambial behaviour with reference to cell length and ring width in Thuja occidentalis L. Can J Bot 38: 177–183

Bannan M W 1960b Ontogenetic trends in conifer cambium with respect to frequency of anticlinal division and cell length. Can J Bot 38: 795–802

Bannan M W 1962 Cambial behaviour with reference to cell length and ring width in Pinus strobus L. Can J Bot 40: 1057–1062

Bannan M W 1963a Cambial behaviour with reference to cell length and ring width in Picea. Can J Bot 41: 811–822

Bannan M W 1963b Tracheid size and rate of anticlinal divisions in the cambium of Cupressus. Can J Bot 41: 1187–1197

Bannan M W 1964a Tracheid size and anticlinal divisions in the cambium of Pseudotsuga. Can J Bot 42: 603–631

Bannan M W 1964b Tracheid size and anticlinal divisions in the cambium of lodgepole pine. Can J Bot 42: 1105–1118

Bannan M W 1965a The rate of elongation of fusiform initials in the cambium of Pinaceae. Can J Bot 43: 429–435

Bannan M W 1965b Ray contacts and rate of anticlinal division in fusiform cambial cells of some Pinaceae. Can J Bot 43: 487–507

Bannan M W 1966a Cell length and rate of anticlinal division in the cambium of the Sequoias. Can J Bot 44: 209–218

Bannan M W 1966b Spiral grain and anticlinal divisions in the cambium of conifers. Can J Bot 44: 1515–1538

Bannan M W 1968a Anticlinal divisions and the organization of conifer cambium. Bot Gaz 129: 107–113

Bannan M W 1968b Polarity in the survival and elongation of fusiform initials in conifer cambium. Can J Bot 46: 1005–1008

Bannan M W, Bayly I L 1956 Cell size and survival in conifer cambium. Can J Bot 34: 769–776

Bannan M W, Whalley B E 1950 The elongation of fusiform cambial cells in Chamaecyparis. Can J Res C 28: 341–355

Barnett J R, Harris J M 1975 Early stages of bordered pit formation in radiata pine. Wood Sci Technol 9: 233–241

Barr N A 1983 Sawing qualities of Eucalyptus saligna S.M. as a guide to selection of trees for breeding and seed collection. Silvicultura 32: 800–801

Baumer M 1952 A propos des figures du bois. Bois For Trop 24: 233–255

Bhat R V, Singh M M 1955 Wrapping papers from chir (Pinus longifolia Roxb) of twisted grain. Indian For 81: 765–773

Birot Y, Arbez M, Azoeuf P, Hoslin R 1979 Variabilité phénotypique de l'angle du fil du bois en fonction de la hauteur chez le Pin laricio et le Douglas. Ann Sci For 36: 165–174

Birot Y, Dufour J, Ferrades P, Azoeuf P, Hoslin R, Tiessier du Cros E 1980 Variabilité de l'angle du fil du bois chez quelques feuillus: Hêtre, Chêne et Eucalyptus dalrympleana. Ann Sci For 37: 19–36

Boas I H 1933 Cross, diagonal and spiral grain in timber. CSIRO Aust For Prod Div Trade Circ 13, 11 pp

Bootle K R 1983 Wood in Australia. McGraw-Hill, Sydney, 443 pp

Bradley C B 1918 The twist of wood fibre in the Tamarac pine. Madrono 1: 63–66

Braun A 1854 Über den schiefen Verlauf der Holzfasern und der dadurch bewirkten Drehung der Bäume. Ber Berlin Akad: 432–484

Brazier J D 1965 An assessment of the incidence and significance of spiral grain in young conifer stems. For Prod J 15: 308–312

Brazier J D 1967 Timber improvement. 1. A study of the variation in wood characteristics in young sitka spruce. Forestry 40: 117–128

Brouse D 1961 Some causes of warping in plywood and veneered products. US For Serv Rep FPL-1952, 8 pp

Brown C L 1964 The influence of external pressure on the differentiation of cells and tissues cultured in vitro. In: Zimmermann M H (ed) The formation of wood in forest trees. Academic Press, New York, London, 562 pp

Brown C L, Sax K 1962 The influence of pressure on the differentiation of secondary tissues. Am J Bot 49: 683–691

Burdon R D 1980 Further results from a genetic survey of Pinus radiata spiral grain and its relationship to other traits. NZ For Serv FRI, Genetics and Tree Improvement Internal Report 188

Burdon R D, Low C B 1981 Wood density and its relationship to other traits in a genetic survey of Pinus radiata. NZ For Serv FRI, Genetics and Tree Improvement Internal Report 205

Burger H 1941 Der Drehwuchs bei den Holzarten 1. Drehwuchs bei Fichte und Tanne. Mitt Schweiz Centralanst Fort Versuchswes 22: 142–163

Burger H 1946 Der Drehwuchs bei Birn- und Apfelbäumen. Schweiz Z Forstwes 97: 119–125

Burger H 1947 Holz Blattmenge und Zuwachs VIII Die Eiche. Mitt Schweiz Centralanst Forst Versuchswes 25: 211–279

Burger H 1950a Holz Blattmenge und Zuwachs X Die Buche. Mitt Schweiz Centralanst Forst Versuchswes 26: 419–468

Burger H 1950b Drehwuchs bei der Lärche. Forstwiss Clb 69: 121–125

Burger H 1953 Das „Arbeiten" imprägnierter Leitungsmasten. Mitt Schweiz Centralanst Forst Versuchswes 29: 177–188

Burley J, Andrew I A 1970 Variation in wood properties of Pinus kesiya Royle ex Gordon (Syn. P. khasya Royle; P. insularis Endlicher): six trees of Assam provenance grown in Zambia. Wood Sci Technol 4:195–212

Burley J, Press A, Morgan J F 1967 Pine wood quality studies in Central Africa 2. Development of laboratory techniques A. Spiral Grain. Proc IUFRO 14 Congr München IX Sect 41: 436–457

Büsgen M, Münch E 1929 The structure and life of forest trees. Transl by Thompson T, Wiley, New York, 436 pp

Butler B T 1931 Twisted trunks of apple trees. Science 73 (1903): 674

Cahn A A 1931 Twisted trees. Science 73 (1899): 561

Canning F 1915 Twisted fibres in chir pine. Indian For 41: 112–116

Champion H G 1924 Contributions towards a knowledge of twisted fibre in trees. Indian For Rec 11: 11–80

Champion H G 1927a An interim report on the progress of investigations into the origins of twisted fibre in Pinus longifolia Roxb. Indian For 53: 18–22

Champion H G 1927b On the twisting of tree trunks. Indian For 53: 236–237

Champion H G 1929 More about spiral grain in conifers. Indian For 55: 57–58

Champion H G 1930 Second interim report on the progress of investigations into the origins of twisted fibre in Pinus longifolia Roxb. Indian For 56: 511–520

Champion H G 1931 Silvicultural research manual for use in India, Vol 1. Govt of India Central Publication Branch, 181 pp

Champion H G 1945 Genetics and forestry. Q J For 32: 74–81

Chattaway M M 1959 Spiral grain in Pinus radiata. CSIRO Aust For Prod Newsletter 250: 3–4

Cheney L S 1896 On the twisting of the grain of pine. Bot Gaz 21: 159–160

Collins J F 1930 On changing the direction of sap conducting tissues. Rhodora 32: 145–146

Copisarow M 1933 The problem of twisted trees. Science 77 (2007): 581–583

Cown D J, McConchie D L 1981 Effects of thinning and fertiliser application on wood properties of Pinus radiata. NZ J For Sci 11: 79–91

Cown D J, McConchie D L, Young G D 1983 Wood properties of Pinus caribaea var. hondurensis grown in Fiji. N Z For Serv Bulletin 17, 54 pp

Cumbie B G 1967 Developmental changes in the cambium of Leitneria floridana. Am J Bot 54: 414–424

Dadswell H F, Fielding J M, Nicholls J W P, Brown A G 1961 Tree-to-tree variations and the gross heritability of wood characteristics of Pinus radiata. Tappi 44: 174–179

Dadswell H F, Nicholls J W P 1959 Assessment of wood qualities for tree breeding 1. In Pinus elliottii var. elliottii from Queensland. CSIRO Aust Div For Prod Tech Pap 4, 16 pp

Das T L, Dayanand T 1983 Amino acid composition in wavy-grained and straight-grained trees of red sanders (Pterocarpus santalinus Linn. F.). Indian J For 6: 158–159

Détienne P 1979 Contrefil a rythme annuel dans les faro, Daniella sp. pl. Rev Bois For Trop 183: 67–71

de Villiers A M 1973 Observations on the timber properties of certain tropical pines grown in South Africa and their improvement by tree breeding. In: Burley J, Nikles D G (eds) Selection and breeding to improve tropical conifers 2: 95–115 de Villiers A M 1974 Observations on the timber properties of certain tropical pines grown in South Africa and their improvement by tree breeding. For S Afr 15: 57–64

Dodd J D 1984 On the shape of cells in the cambial zone of Pinus sylvestris L. Am J Bot 35: 666–682

Edwards L N 1931 Early utilitarian application of twist in trees. Science 73: 674–675

Egierszdorff S, Tomaszewski M 1973 Auxin-dependent accumulation of photosynthates in cambium and wood formation in Scots pine. Trans 3rd symposium on accumulation and translocation of nutrients and regulators in plant organisms

Elliott G K 1958 Spiral grain in second-growth Douglas fir and western hemlock. For Prod J 8: 205–211

Elliott G K 1967 Some problems of spiral grain with special reference to conifers. Proc IUFRO 14 Congr München IX Sect 41: 413–435

Erdesi J 1957 (Spiral grain in ash). Sumarstvo 10 (5/6): 345–350

Etholén K, Huuri L 1982 Viskoivua kätsitteleva kirjallisuus. [Bibliography on curly birch, Betula pendual var. carelica (Mercklin)]. Fol For 502, 24 pp

Fielding J M 1953 Variations in Monterey pine. Aust For Timber Bur Bull 31, 43 pp

Fielding J M 1967 Spiral grain in Pinus radiata plantations in the Australian Capital Territory. Timber Bur Aust Leafl 103, 28 pp

Foulger A N 1966 A spiral grain – fibril angle relationship in coniferous stems. US For Serv Rep FPL – unnumbered, 10 pp

Foulger A N 1969 Through-bark measurements of grain direction: preliminary results. For Sci 15: 92–94

Gerischer G F R, de Villiers A M 1963 The effect of heavy pruning on timber properties. For S Afr 3: 15–41

Gerischer G F R, Kromhout C P 1964 Note on breast height spirality in dominant trees of Pinus patula, Pinus taeda and Pinus elliottii with special reference to tree breeding. For S Afr 5: 81–97

Gersani M 1985 Appearance of new transport capacity in wounded plants. J Exp Bot 36: 1809–1816

Goldsmith M H M 1969 Transport of plant growth regulators. In: Wilkins M B (ed) Physiology of plant growth and development. McGraw-Hill, London, 695 pp

Gorrie R M 1938 Twisted fibre in Pinus excelsa (blue pine). Indian For 64: 185–188

Greenidge K N H 1955 Observations on the movement of moisture in large woody stems. Can J Bot 33: 202–221

Greenidge K N H 1958 Rates and patterns of moisture movement in trees. In: Thimann K V (ed) The physiology of forest trees. The Ronald Press, New York, 19–41

Griffith A L 1946 Twisted fibre in conifers. Indian For 72: 512–513

Gunther A E 1929 The twisting of the Tamarac pine. Scott For J 43: 81–94

Harding K J 1986 Status report on wood quality/tree breeding cooperative research projects (1983–1985) of Queensland Department of Forestry. Proc Aust For Council, Res Working Group 1, Forest Genetics, Tasmania 1986

Harris J M 1965 Preliminary studies of spiral grain in radiata pine. Proc IUFRO Melbourne Sect 41, Vol 1, 23 pp

Harris J M 1967 The causes of spiral grain in the corewood of radiata pine. Proc IUFRO 14 Congr München IX Sect 41: 363–383

Harris J M 1969 On the causes of spiral grain in corewood of radiata pine. NZ J Bot 7: 189–213

Harris J M 1973 Spiral grain and xylem polarity in radiata pine: microscopy of cambial reorientation. NZ J For Sci 3: 363–378

Harris J M 1981 Spiral grain formation. In: Barnett J R (ed) Xylem cell development. Castle House, Tunbridge Wells, 256–272

Harris J M 1984 Non-destructive assessment of spiral grain in standing trees. NZ J For Sci 14: 395–399

Harris J M, McConchie D L 1978 Wood properties of Pinus radiata infected with Dothistroma pini. NZ J For Sci 8: 410–416

Harris J M, McConchie D L, Povey W A 1979 Wood properties of clonal radiata pine grown in soils with different levels of available nitrogen, phosphorus and water. NZ J For Sci 8: 417–430

Hartig R 1895 Über den Drehwuchs der Kiefer. Forst – Naturwiss Z 4: 313–326

Hartig T 1858 Über die Bewegung des Saftes in den Holzpflanzen. Bot Ztg 16: 329–335, 337–342

Harzmann L-J 1965 Mitteilung über das Auftreten des Wecheldrehwuchses bei Liquidambar formosana. Arch Forstwes 14: 801–806

Haskins C P, Moore C N 1933 The physiological basis of the twisting habit in plant growth. Science 77: 283
Heikinheimo O 1951 Experiments in growing curly birch. Commun Inst For Fenn 39.5: 24–26
Hejnowicz Z 1961 Anticlinal division, intrusive growth, and loss of fusiform initials in nonstoried cambium. Acta Soc Bot Pol 30: 729–748
Hejnowicz Z 1964 Orientation of the partition in pseudotransverse division in cambia of some conifers. Can J Bot 42: 1685–1691
Hejnowicz Z 1967a Interrelationship between mean length, rate of intrusive elongation, frequency of anticlinal divisions and survival of fusiform initials in cambium. Acta Soc Bot Pol 36: 367–378
Hejnowicz Z 1967b Changes in anatomy and physiology of the cambium as related to spiral grain development. Proc IUFRO 14 Congr München IX Sect 41: 352–362
Hejnowicz Z 1968 The structural mechanism involved in the changes of grain in timber. Acta Soc Bot Pol 37: 347–365
Hejnowicz Z 1971 Upward movement of the domain pattern in the cambium producing wavy grain in Picea excelsa. Acta Soc Bot Pol 40: 499–512
Hejnowicz Z 1972 The orientation of the plane of cell division in cambia of trees and factors controlling it. Univ Wroclaw Tech Rep April 1967–April 1972, 73 pp
Hejnowicz Z 1973a Domain pattern in cambium and formation of spiral grain. Proc IUFRO Div 5 S Afr, 2: 459–466
Hejnowicz Z 1973b Morphogenetic waves in cambia of trees. Plant Sci Lett 1: 359–366
Hejnowicz Z 1974 Pulsations of domain length as support for the hypothesis of morphogenetic waves in the cambium. Acta Soc Bot Pol 43: 261–271
Hejnowicz Z 1975 A model for morphogenetic map and clock. J Theor Biol 54: 345–362
Hejnowicz Z 1980 Tensional stress in the cambium and its developmental significance. Am J Bot 67: 1–5
Hejnowicz Z, Brański S 1966 Quantitative analysis of cambium growth in Thuja. Acta Soc Bot Pol 35: 395–400
Hejnowicz Z, Krawczyszyn J 1969 Oriented morphogenetic phenomena in cambium of broadleaved trees. Acta Soc Bot Pol 38: 547–560
Hejnowicz Z, Romberger J A 1973 Migrating cambial domains and the origin of wavy grain in xylem of broadleaved trees. Am J Bot 60: 209–222
Hejnowicz Z, Romberger J A 1979 The common basis for wood grain figures is the systematically changing orientation of cambium fusiform cells. Wood Sci Technol 14: 89–96
Hejnowicz Z, Zagórska-Marek B 1974 Mechanism of changes in grain inclination in wood produced by storeyed cambium. Acta Soc Bot Pol 43: 381–398
Hellawell C R 1967 Some practical background to considerations of the economic significance of spiral grain. Proc IUFRO 14 Congr München IX Sect 41: 458–469
Hendrickson W H, Vité J P 1960 The pattern of water conduction and tracheidal alignment in Douglas fir. Contrib Boyce Thompson Inst 20: 353–361
Hepler P K 1981 Morphogenesis of tracheary elements and guard cells. In: Kiermayer O (ed) Cell biology monographs Vol 8 Cytomorphogenesis in plants. Springer, Berlin, Heidelberg, New York, 439 pp
Herrick E H 1932 Further note on twisted trees. Science 76 (1932): 406–407
Houkal D 1982 Spiral grain in Pinus oocarpa. Wood Fiber 14: 320–330
Howard N F 1932 Twisted trees. Science 77 (1935): 132–133
Hughes J F 1973 The wood structure of Pinus caribaea Morelet in relation to use characteristics, growth conditions and tree improvement. In: Burley J, Nikles D G (eds) Selection and breeding to improve some tropical conifers 2: 13–21
Hughes J F, Andrew I A 1974 New techniques for the evaluation of wood quality and their application in schemes for planting fast growing species in the tropics. 10th Commonw For Conf, Oxford GB, 16 pp
Jablanczy A 1969 Spiral growth in juvenile spruce seedling. For Chron 45: 44
Jacobs M R 1935 The occurrence and importance of spiral grain in Pinus radiata in the Federal Capital Territory. Aust Commonw For Bur Leafl 50, 5 pp
Jacot A P 1931 Tree twist. Science 74 (1927): 567
James W L, Yen Y-H, King R J 1985 A microwave method for measuring moisture content, density, and grain angle of wood. US For Serv Res Note FPL-0250, 9 pp

Jane F W 1956 The structure of wood. Adam and Charles Black, London, 427 pp

Johnsson H 1950 Avkommor av masurbjörk. (Offspring of curly birch). Årsberätt Fören Växtför Skogsträd 18–29

Jones A T 1931 Trees with twisted bark. Science 73 (1891): 341

Jones B E 1963 Cell adjustments accompanying the development of spiral grain in a specimen of Pseudotsuga taxifolia Britt. Commonw For Rev 43: 151–158

Kaasa J 1976 Vridd vekst hos gran og furu. (Spiral grain in fir and pine). Tidskr Skogbruk 84: 299–309

Kadambi K 1951 On the nature of twisted fibre and the occurrence of interlocked fibre in some trees. 8th Silvic Conf Dehra Dun India 227–228

Kadambi K, Dabral S N 1955 On twist in chir (Pinus longifolia Roxb.). Indian For 81: 58–64

Kano T, Nakagawa S, Saito H, Oda S 1964 [On the quality of larch timber. (Larix leptolepis Gordon)]. Report 1. Bull Govt For Exp Stn Tokyo 162: 1–44

Kano T, Nakagawa S, Saito S, Oda S, Shigematsu Y 1965 [On the quality of larch timber (Larix leptolepis Gordon)]. Report 2. Bull Govt For Exp Stn (Tokyo) 182: 113–147

Kano T, Saito H 1965 (A new device for measurement of fiber direction). Mokuzai Gakkaishi 11: 33–35

Kawaguchi N, Yamamoto H, Takahashi M, Takizawa T 1979 (The spiral grain of karamatsu, Larix leptolepis Gord., growing in a thinned forest). J Hokkaido For Prod Res Inst 328: 4–8

Kawaguchi N, Yamamoto H, Takahashi M, Takizawa T 1982 (On the spiral grain in the stem of Japanese larch). Proc Japan Wood Res Soc Hokkaido 10: 4–7

Keating W G 1966 Twisting of Corsican and radiata pine telephone poles. Aust For 30: 184–190

Kedharnath S 1963 Present status of forest tree breeding in India. Proc FAO World Consult For Genet Stockholm FAO-FORGEN 63/-7/6, 10 pp

Kedharnath S, Rawat M S 1976 Studies on variation in fibre morphology in wavy grained and straight grained trees of red sanders. Indian For 102: 441–446

Kedharnath S, Rawat M S, Uniyal D P, Laksumikantham D 1976 Studies on field grafting and the growth of the grafts in red sanders. Indian For 102: 761–765

Keller R 1973 Caractéristiques du bois de pin maritime variabilité et transmission héréditaire. Ann Sci For 30: 31–62

Keller R, Azoeuf P, Hoslin R 1974 Détermination de l'angle de la fibre torse d'arbres sur pied à l'aide d'un traceur radioactif. Ann Sci For 31: 161–169

Kennedy R W, Elliott G K 1957 Spiral grain in red alder. For Chron 33: 238–257

Kingston R S T 1947 The variation of tensile strength and modulus of elasticity of hoop pine veneer with the direction of the grain. J Counc Sci Ind Res Aust 20: 338–345

Kirschner H, Sachs T, Fahn A 1971 Secondary xylem reorientation as a special case of vascular tissue differentiation. Israel J Bot 20: 184–198

Kloot N H, Page M W 1959 A study of distortion in radiata pine scantlings. CSIRO Aust Div For Prod Tech Pap 7, 24 pp

Kloot N H, Schuster K B 1958 The effect of cross-grain on the bending properties of jarrah scantlings. Aust J Appl Sci 9: 9–17

Knigge W, Schulz H 1959 Methodische Untersuchungen über die Möglichkeit der Drehwuchsfeststellung in verschiedenen Alterszonen von Laubhölzern. Holz Roh-Werkst 17: 341–351

Knight T A 1807 On the formation of the bark of trees. Royal Soc London Philos Trans 103–113

Knorr F 1932 What causes twisted trees? J Hered 23: 49–52

Koehler A 1924 The properties and uses of wood. McGraw-Hill, New York, 354 pp

Koehler A 1931 More about twisted grain. Science 73 (1896): 477

Koehler A 1939 Heredity versus environment in improving wood in forest trees. J For 37: 683–687

Koehler A 1949 Growing better timber. In Trees. US Dept Agric Yearbook: 200–205

Koehler A 1955 Guide to determining slope of grain in lumber and veneer. US For Serv Rep FPL-1585, 58 pp

Kohl E J 1933 An explanation of the cause of spiral grain in trees. Science 73 (2012): 58–59

Kollmann F F P, Côté W A 1968 Principles of wood science and technology 1. Solid Wood. Springer, Berlin, Heidelberg, New York, 592 pp

König E 1947 Spiraliger Verlauf der Holzfaser (Drehwuchs). Holz-Zbl 23: 183–185

Kostov P, Todorov S 1977 (Wood defects in oak logs). Gorsko Stoparstvo 33(a): 35–40

Kozlowski T T 1961 The movement of water in trees. For Sci 7: 178–192

Kozlowski T T, Hughes J F, Leyton L 1967 Movement of injected dyes in gymnosperm stems in relation to tracheid alignment. Forestry 40: 207–219

Kozlowski T T, Leyton L, Hughes J F 1965 Pathways of water movement in young conifers. Nature 205: 830

Kozlowski T T, Winget C H 1963 Patterns of water movement in forest trees. Bot Gaz 124: 301–311

Krahl-Urban J 1953a Hinweise auf individuelle Erbanlagen bei Eichen und Buchen. Z Forstgenet 2: 51–59

Krahl-Urban J 1953b Drehwuchs bei Eichen und Buchen. Allg Forstz 8: 540–542

Krahl-Urban J 1960 Einige Ergebnisse von Drehwuchs – Untersuchungen bei Rotbuchen. Silvae Genet 9: 139

Krahl-Urban J 1961 Über den Drehwuchs bei Buchen. Forstarchiv 32: 197–201

Krahl-Urban J 1967 Über den Drehwuchs bei Buchen. Proc IUFRO 14 Congr München IX Sect 41: 384–397

Krawczyszyn J 1971 Unidirectional splitting and uniting of rays in the cambium of Platanus accompanying the formation of interlocked grain in wood. Acta Soc Bot Pol 40: 57–79

Krawczyszyn J 1972 Movement of the cambial domain pattern and mechanism of formation of interlocked grain in Platanus. Acta Soc Bot Pol 41: 443–461

Krawczyszyn J 1973 Domain pattern in the cambium of young Platanus stems. Acta Soc Bot Pol 42: 637–648

Krawczyszyn J 1977 Cambial domain pattern in the root and root collar of Platanus acerifolia. Acta Soc Bot Pol 46: 531–541

Krawczyszyn J, Romberger J A 1975 Wood grain patterns allow study of the time aspect of slowly migrating morphogenic waves in cambia of trees. Chronobiologia 2 Supplement 1: 38–39

Krawczyszyn J, Romberger J A 1979 Cyclical cell length changes in wood in relation to storied structure and interlocked grain. Can J Bot 57: 787–794

Krawczyszyn J, Romberger J A 1980 Interlocked grain, cambial domains, endogenous rhythms, and time relations, with emphasis on Nyssa sylvatica. Am J Bot 67: 228–236

Krempl H 1965a Spiral growth in the log and branchwood of spruce. Proc IUFRO Melbourne Sect 41, Vol 1, 10 pp

Krempl H 1965b Preliminary research results on spiral grain in Norway spruce. Proc IUFRO Melbourne, Sect 41, Vol 1, 17 pp

Krempl H 1965c Drehwuchs in Stamm- und Astholz bei der Fichte. Holzforsch Holzverwert 17: 21–26

Krempl H 1970 Untersuchungen über den Drehwuchs bei Fichten. Mitt Forst Bundes-Versuchsanst, Wien, 118 pp

Krogh P M D 1952 The twisting of wooden telephone poles in service in South Africa. 6th Brit Commonw For Conf, Canada, 15 pp

Kromhout C P 1966 A note on spirality measurement on stem samples for tree breeding purposes. For S Afr 6: 79–85

Kromhout C P, Bosman D L 1981 The influence of short rotation forestry on wood production for sawn wood and veneer. Proc IUFRO 17 Congr Japan Div 5: 60–75

Kromhout C P, Toon R E 1977 Variation of wood properties of some tropical species grown in plantations in southern Africa. In: Nikles D G, Burley J, Barnes R D (eds) Progress and problems of genetic improvement of tropical forest trees 1: 8–20

Langlands I 1932 Tests on samples of some Tasmanian woods. Aust Div For Prod Project TM 7–3, 4 pp

Langlands I 1933 The effect of the size of specimen on the toughness of interlocked karri. Aust Div For Prod Project TM 7–3, 8 pp

Langlands I 1936 A discussion of special tests on the compressive strength of green karri (Eucalyptus diversicolor). Aust Div For Prod Tech Pap 19, 31 pp

Lasschuit J A 1951 Alteration in spirally curved trees of Pinus merkusii. Tectona 41: 179–188

Liese W, Ammer U 1962 Anatomische Untersuchungen an extrem drehwüchsigem Kiefernholz. Holz Roh-Werkst 20: 339–346

Limaye V D 1954 Interlocking of grain in Indian timbers. Indian For 80: 6–9

Lintilhac P M, Vesecky T B 1984 Stress-induced alignment of division plane in plant tissues grown in vitro. Nature 307: 363–364

Lloyd E 1984 Plants under strain. Nature 307: 319–320

Löhr E 1966 Drehwuchs bei Rosskastanie, Eiche und anderen Bäumen. Bot Tidskr 62: 337–347

Lowery D P 1966 A spiral grain classification system and its application. For Prod J 16: 47–50

Lowery D P 1967a Some studies of spiral grain. Proc IUFRO 14 Congr München IX Sect 41: 470–483

Lowery D P 1967b Spiral grain in individual growth rings. J For 65: 120–121

Lowery D P, Erickson E C O 1967 The effect of spiral grain on pole twist and bending strength. US For Serv Res Pap INT-35, 16 pp

Luo L-C, Xu L-F 1983 Effects of cross grain on the physical and mechanical properties of the wood of Yunnan pine. Sci Silvae Sin 19: 382–389

Lutz H J 1943 Injuries to trees caused by Celastrus and Vitis. Bull Torrey Bot Club 70: 436–439

MacDaniels L H, Curtis O F 1930 The effect of spiral ringing on solute translocation and the structure of the regenerated tissues of the apple. Cornell Univ Agri Expt Stn Memoir 133, 31 pp

Markwardt L J, Wood L W 1958 Simplified principles for structural grading of timber. US For Serv Rep FPL-2112, 20 pp

Martley J F 1920 Double cross-grain. Ann Appl Biol 7: 224–268

Mayer-Wegelin H 1956 Die biologische, technologische und forstliche Bedeutung des Drehwuchses der Waldbäume. Forstarchiv 27: 265–271

McBride C F 1967 Some economic losses in producing lumber from spiral grained logs. Can For Serv Inf Rep VP-X-17, 12 pp

McBride C F 1969 Spiral grain losses cost B.C. industry $5 million a year. BC Lumberman 53: 46–50

McCarthy E F, Hoyle R J 1918 Knot zones and spiral in Adirondack red spruce. J For 16: 777–791

McKinney H H, Sando W J 1934 Twisted wheat and twisted trees. J Hered 25: 261–263

McLauchlan T A, Norton J A, Kusec D J 1973 Slope of grain indicator. For Prod J 23(5): 50–55

Meyer K A 1949 Sprachliche und literarische Bemerkungen zum Problem „Drehwuchs". Mitt Schweiz Anst Forst Versuchswes 26: 331–347

Meylan B A, Butterfield B G 1978 Helical orientation of the microfibrils in tracheids, fibres and vessels. Wood Sci Techol 12: 219–222

Mikami S 1973 Preliminary studies on the genetic improvement in spiral grain in Japanese larch. Proc IUFRO Div 5 S Afr, 11 pp

Mikami S, Nagasaka K 1974 (Geographic variation in spiral grain of Larix leptolepis Gord.). J Japan For Soc 56: 228–230

Mikami S, Nagasaka K 1975 (Selection for minimising spiral grain in Larix leptolepis Gord.). Bull Govt For Expt Stn Meguro 276: 1–22

Mikami S, Watanabe M, Ohta N 1972 (Clonal variation in spiral grain of Larix leptolepis Gord.). J Japan For Soc 54: 213–217

Miller E C L 1934 Spiraling in trees. Science 80: 35

Misra P 1939 Observation on spiral grain in the wood of Pinus longifolia Roxb. Forestry 13: 118–133

Misra P 1943 Correlation between eccentricity and spiral grain in the wood of Pinus longifolia. Forestry 17: 67–80

Miyaki M, Hatakeyama S, Goshu K 1983 (Relationship between grain inclination and axis of pith when using the splitting method). J Japan For Soc 65: 183–186

Moll F 1947 Spiraliger Verlauf der Holzfaser (Drehwuchs). Holz-Zbl 27: 221

Nakagawa S 1972 (Distribution of spiral grain within stem and the spirality pattern on Larix leptolepis Gordon). Bull Govt For Expt Stn Meguro 248: 97–120

Nakagawa S 1980 (Relationship of spiral grain between mother trees and clones of Larix leptolepis Gord.). Bull Japan For Prod Res Inst 312: 21–43

Nakagawa S 1986 Variation in spiral grain of Akamatsu (Pinus densiflora Siel et Zucc). Bull For and For Prod Res Inst 335: 15–30

Neeff F 1914 Über Zellumlagerung. Ein Beitrag zur experimentellen Anatomie. Z Bot 6: 465–547

Newall R J, Gardiner A S 1963 Bark form and wood figure in home-grown birch. UK DSIR and For Comm Special Rept 18, 26 pp

Newman I V 1956 Miscellaneous notes on inclination of grain in Pinus radiata D. Don and Pinus Sp. CSIRO Aust Div For Prod Lab Rept 1 Project W17, 16 pp

Nicholls J W P 1963 The relation of spiral grain to wood quality. Proc IUFRO Madison Sect 41, 25 pp

Nicholls JWP 1965 The possible causes of spiral grain. Proc IUFRO Melbourne Sect 41, Vol I, 7 pp

Nicholls J W P 1967a Preliminary observations on the change with age of the heritability of certain wood characters in Pinus radiata clones. Silvae Genet 16: 18–20

Nicholls J W P 1967b Assessment of wood qualities for tree breeding. IV Pinus pinaster Ait. grown in Western Australia. Silvae Genet 16: 21–28

Nicholls J W P 1971a Preferred methods for the investigation of spiral grain. Proc IUFRO 15 Congr, 5 pp

Nicholls J W P 1971b Some aspects of the genetic and environmental control of spiral grain in Pinus species. Proc IUFRO 15 Congr, 13 pp

Nicholls J W P 1971c The effect of environmental factors on wood characteristics. 2. The effect of thinning and fertilizer treatment on the wood of Pinus pinaster. Silvae Genet 20: 67–73

Nicholls J W P 1973 Is fast grown Pinus radiata acceptable from a utilisation viewpoint? Proc 3 Meet Res Comm Aust For Council, 4 pp

Nicholls J W P, Brown A G 1971 The ortet-ramet relationship in wood characteristics of Pinus radiata. Appita 25: 200–209

Nicholls J W P, Brown A G 1972 Wood characteristics of colchicine-treated Pinus radiata. J Inst Wood Sci 6: 13–16

Nicholls J W P, Brown A G 1974 The relationship between ring width and wood characteristics in double-stemmed trees of radiata pine. NZ J For Sci 4: 105–111

Nicholls J W P, Brown A G 1976 The effect of hedging on wood characteristics of Pinus radiata. NZ J For Sci 6: 397–408

Nicholls J W P, Brown A G, Pederick L A 1974 Wood characteristics of sexually and vegetatively produced Pinus radiata. Aust J Bot 22: 19–27

Nicholls J W P, Dadswell H E, Fielding J M 1964 The heritability of wood characteristics of Pinus radiata. Silvae Genet 13: 68–71

Nicholls J W P, Eldridge K 1980 Variation in some wood and bark characteristics in provenances of Pinus radiata D. don. Aust For Res 10: 321–335

Nicholls J W P, Fielding J M 1965 The effect of growth rate on wood characteristics. Appita 19: 24–30

Nicholls J W P, Pawsey C K, Brown A G 1976 Further studies on the ortet-ramet relationship in wood characteristics of Pinus radiata. Silvae Genet 25: 73–79

Nicholls J W P, Pederick L A, Brown A G 1977 A summary of the ortet-ramet relationship in wood characteristics of Pinus radiata. Appita 30: 496–502

Nicholls J W P, Waring H D 1977 The effect of environmental factors on wood characteristics IV. Irrigation and partial droughting of Pinus radiata. Silvae Genet 26: 107–111

Nikles D G, Smith W J 1974 Wood studies. Progress Rept Queensland Dep For 60–62

Nix L E, Wodzicki T J 1974 The radial distribution and metabolism of IAA-^{14}C in Pinus echinata stems in relation to wood formation. Can J Bot 52: 1349–1355

Northcott P L 1957 Is spiral grain the normal growth pattern? For Chron 33: 335–352

Northcott P L 1958 Spiral grain in wood. BC Lumberman Nov, 2 pp

Northcott P L 1959 Use of spiral grained wood. BC Lumbermn Feb, 3 pp

Northcott P L 1965 The effects of spiral grain on the usefulness of wood. Proc IUFRO Melbourne Sect 41, Vol 1, 15 pp

Noskowiak A F 1959 Spiral grain patterns in red pine and relationship of age and radial growth rate to change of grain angle. Ph D Thesis, SUNY Coll For, Syracuse NY, 273 pp

Noskowiak A F 1963 Spiral grain in trees: a review. For Prod J 13: 266–275

Noskowiak A F 1968 Spiral grain patterns from increment cores. For Prod J 18(4): 57–60

Ohkano T 1984 Estimate of spiral grain in the stem of standing tree by the branches. Proc Pacific Regional Wood Anat Conf Tsukuba Japan, 43–45

Ohkura S 1958 (On the macroscopic features of twisted fibre in trees). J Fac Agri Shinshu Univ 8: 59–100

Ohkura S, Ozawa K, Takashima K, Takeiri K 1961 (On the twisting warp of wood). Mokuzai Gakkaishi 7: 205–207

Onodera S, Takahashi M, Kawaguchi N 1977 (Quality examination of planted tree grown in Hokkaido. II Larch species and its F1 hybrids). Rep Hokkaido For Prod Res Inst 66: 32–110

Ozawa K 1972 [Features in spiral grain of karamatsu (Larix kaempferi Sarg.) from a stand in Tohoku district]. J Japan For Soc 54: 269–274

Ozawa K 1973 [Features in spiral grain of karamatsu (Larix leptolepis Gord.) from a stand in Chubu district]. J Japan For Soc 55: 221–226

Paterson D N 1967 Spiral grain in East African exotic softwood plantations E Afr Agr For J 33: 286–291

Paul B H 1955 Importance of wood quality in tree breeding. J For 53: 659–661

Paul B H 1956 Changes in spiral grain direction in ponderosa pine. US For Serv Rep FPL-2058, 4 pp

Paul B H 1957 Lengthwise shrinkage in ponderosa pine. For Prod J 7: 408–410

Pawsey C K 1965 A study of spiral grain in clones of Pinus radiata. Aust For 29: 89–95

Pederick L A 1970 Selection criteria. Proc Res Comm Aust For Council, Appendix 10, 4 pp

Pederick L A 1971 Inheritance of spiral grain in young radiata pine. Proc IUFRO 15 Congr, 9 pp

Pederick L A 1973 Plus tree register. Proc Res Comm Aust For Council Appendix 12, 8 pp

Pelle D N 1967 A special style for spiral grain estimation of Pinus radiata. Aust For Res 3: 54

Pemberton C C 1929 A revolving growth movement in conifers. Univ Washington For Club Quarterly 8(4): 5–9

Persson A 1954 Plantagefrö av Masurbjörk. (Seed orchard production of figured birch). Skogen 41: 160–163

Phillips E W J 1963 Timber improvement by tree selection and breeding. Proc World Consult For Genet Tree Improv, Stockholm, FAO-FORGEN 63, Sect 7, Vol II, 5 pp

Phillips G E, Bodig J, Goodman J R 1981 Flow-grain analogy. Wood Sci 14: 55–64

Philpson W R, Ward J M, Butterfield B G 1971 The vascular cambium. Chapman and Hall, London, 182 pp

Pillow M Y 1955 Detection of figured wood in standing trees. US For Serv Rep FPL-2034, 10 pp

Preston R D 1949 Spiral structure and spiral growth: the development of spiral grain in conifers. Forestry 23: 48–55

Priestley J H 1930 Studies in the physiology of cambial activity. II. The concept of sliding growth. New Phytol 29: 96–140

Priestley J H 1945 Observations on spiral grain in timber. Am J Bot 32: 277–284

Pyszyński W 1972 Downward movement of the domain pattern in Aesculus cambium producing wavy-grained xylem. Acta Soc Bot Pol 41: 511–517

Pyszyński W 1977a Complex wavy grain in the stem of Aesculus. Acta Soc Bot Pol 46: 231–249

Pyszyński W 1977b Mechanism of formation of spiral grain in Aesculus stems: dissymmetry of deformation of stems caused by cyclic torsion. Acta Soc Bot Pol 46: 501–522

Quirk J T, Smith D M, Freese F 1975 Effects of mechanical stress on growth and anatomical structure of red pine (Pinus resinosa Ait): torque stress. Can J For Res 5: 691–699

Quirk J T, Smith D M, Freese F, Wolter K E 1972 Stimulatory effect of torque on the radial growth of red pine trees. Plant Physiol 49: Suppl, p 5

Rao H S 1954 The phenomenon of twisted trees. Indian For 80: 165–170

Rault J P, Marsh E K 1952 The incidence and silvicultural implications of spiral grain in Pinus longifolia Roxb. in South Africa and its effect on converted timber. Proc Commonw For Conf Canada, For Prod Inst, Pretoria, 21 pp

Raunecker H 1957 Beobachtungen über Drehwuchs der Kiefer. Holz-Zbl 83: 1221

Richens R H 1945 Forest tree breeding and genetics. Imp Agr Bur Joint Publ 8, 79 pp

Ruden T 1954 Om valbjørk og endel andre unormale veddannelser hos bjørk [On speckled birch ("mazer-birch") and some other forms of curled birch]. Norske Skogforsøksvesen 12(43): 451–505

Rudinsky J A, Vité J P 1959 Certain ecological and phylogenetic aspects of the pattern of water conduction in conifers. For Sci 5: 259–266

Sachs T 1981 Polarity changes and tissue organisation in plants. In: Schweiger H G (ed) Cell biology 1980–1981. Springer, Berlin, Heidelberg, New York, 489–496

Sachsse H 1965 The effect of the rate of growth on the occurrence of spiral grain. Proc IUFRO Melbourne Sect 41, 15 pp

Samson M 1984 Measuring general slope of grain with the slope-of-grain indicator. For Prod J 34: 27–32

Sanyal S L, Gulati A S, Jain J D 1980 A note on the twisted chir poles for overhead power and communication lines. J Timber Dev Assoc (India) 26: 5–8

Savidge R A 1985 Prospects for manipulating vascular-cambium productivity and xylem-cell differentiation. In: Cannell M G R, Jackson J E (eds) Attributes of trees as crop plants. Institute of Terrestrial Ecology-GB 208–227

Savidge R A, Farrar J L 1984 Cellular adjustments in the vascular cambium leading to spiral grain formation in conifers. Can J Bot 62: 2872–2879

Savidge R A, Wareing P F 1981 Plant-growth regulators and the differentiation of vascular elements. In: Barnett J R (ed) Xylem cell development. Castle House, Tunbridge Wells, 192–235

Sax K, Dickson A Q 1956 Phloem polarity in bark regeneration. J Arnold Arbor 37: 173–179

Scheiber C 1973 Phänomen Wechseldrehwuchs. Holzindustrie 26: 151–155

Schmidt J D K, Smith W J 1961 Wood quality evaluation and improvement in Pinus caribaea Morelet. Queensland For Serv Res Note 15, 66 pp

Schmucker T 1924 Rechts- und Linkstendenz bei Pflanzen. Beih Bot Zbl 41: 51–81

Schmucker T 1956 Forstgenetik: einige Befunde am Rande. Forstwiss Cbl 75: 32–41

Schreiber M 1926 Beiträge zur Kenntnis der Drehwüchsigkeit der Bäume, von H G Champion. Cbl Ges Forstwes 52: 45–59

Schreiner E J 1958 Possibilities for genetic improvement in the utilisation potentials of forest trees. Silvae Genet 7: 122–128

Sears P M 1950 The mystery of twisted trees. Nat History 59: 469–473

Seifriz W 1933a Twisted trees and the spiral habit. Science 77 (1985): 50–51

Seifriz W 1933b More about the spiral habit. Science 78 (2025): 361–363

Sen N N 1955 Letter to the editor. Indian For 81:153

Sinnott E W 1960 Plant morphogenesis. McGraw-Hill, New York, 550 pp

Sinnott E W, Bloch R 1939 Changes in intercellular relationship during the growth and differentiation of living plant tissues. Am J Bot 26: 625–634

Small J K 1932 The crux of twisted and contorted tree trunks. J NY Bot Gard 33: 189–195

Smith D M, Wahlgren H E, Genston G W 1971 Effect of irrigation and fertilization on wood quality of young slash pine. Proc API/TAPPI Symposium FPL-Madison WI, D1–14

Smith W J 1970 Improvement in yield and wood quality through selection criteria and indices. Proc Res Comm Aust For Council, 29 pp

Smith W J 1973 Wood yield, properties and quality in Queensland-grown Pinus caribaea Morelet. In: Burley J, Nikles D G (eds) Selection and breeding to improve some tropical conifers 2: 45–69

Smith W J, Eccles D B 1981 Comparative wood quality evaluation of "foxtail" and "normal" trees in Pinus caribaea var. hondurensis. Proc 20th For Prod Conf Melbourne 1981, 8 pp

Smith W J, Smart J A J 1972 Factors affecting wood quality in softwood plantations. Paper Aust TIS Conf, Queensland Aust

Smythies E A 1915 Notes on the twisted fibre in chir pine. Indian For 41: 69–75

Steeves T A, Sussex I M 1972 Patterns in plant development. Prentice-Hall, New Jersey, 302 pp

Stern K 1969 Minimum standards for provenance testing and progeny testing for certification purposes. Sec FAO IUFRO World Consult For Tree Breed Washington DC 2: 1448–1451

Stevens W C, Johnston D D 1960 Distortion caused by spiralled grain. Timb Technol 68(2252): 217–218

Sweet G B, Harris J M 1976 Wood properties of Pinus radiata: seed-grown trees compared with grafts from different-aged ortets. NZ J For Sci 6: 114–121

Takemura T 1978 The distortion of the flat-sawn board with alternating spiral grain by drying. Mokuzai Gakkaishi 24: 183–189

Takizawa T, Kawaguchi N, Takahashi M 1981 (The relationship between the spiral grain and the twist of sawn larch squares containing the pith). J Hokkaido For Prod Res Inst 357: 7–11

Teissier du Cros E, Kleinschmit J, Azoeuf P, Hoslin R 1980 Spiral grain in beech, variability and heredity. Silvae Genet 29: 5–13

Teodoresco E C, Popesco C T 1915 Sur le tissu libérien et son rôle dans la circulation des substances organiques chez les végétaux supérieurs. Ann Sci Univ Jassy 9: 215–242

Thair B W 1968 Cambial polarity as revealed by grafting experiments. Thesis Univ Saskatchewan Saskatoon

Thair B W, Steeves T A 1976 Response of the vascular cambium to reorientation in patch grafts. Can J Bot 54: 361–373

Thunell B 1951 Über die Drehwüchsigkeit. Holz Roh-Werkst 8: 293–297

Tiemann H D 1941 What is "spiral grain"? South Lumberman 163: 55–57

Timell T E 1986 Compression wood in gymnosperms. Springer, Berlin, Heidelberg, New York, Tokyo, 2150 pp

Toda M, Mikami S 1976 The provenance trials of Japanese larch established in Japan and the tentative achievements. Silvae Genet 25: 209–216

Toker R 1952 Türkiyede Okaliptus (E. rostrata) ün maden diregi bakimindan teknik özellikleri hakkinda arastirmalar. [Studies of the properties of eucalyptus wood (E. rostrata) with respect to its use as mine timber]. Orman Fakültesi Dergisi, Istanbul Univ 2: 126–152

Toumey J W, Korstian C F 1942 Seeding and planting in the practice of forestry. Wiley, New York, 520 pp

Troup R S 1916 Pinus longifolia Roxb. A silvicultural study. Indian For Mem 1: 1–126

Turnbull J M 1942 A preliminary investigation into the causes of torsion in pine stems. J S Afr For Assoc 8: 112–117

Vaclav E 1969 The quality of wood of the technical forms of birch. Proc Second World Consult For Tree Breed 1: 437–446

Van der Merve J J 1973 Strength and warp in kiln-dried South African pine timbers as affected by the drying process, spiral grain and other factors. CSIR S Afr Special Rept Hout 56: 10 pp

Van Iterson G 1943 Modellen en theorieën ter toelichtung van het mechanisme van der spiraalvormigen groei. (Models and theories to illustrate the mechanism of spiral growth). Nederland Akad Wetenschappen 52: 58–68 Seen as CSIRO Aust Transl 1784

Van Oye P 1926 Sur la torsion des troncs d'arbres. Bull Soc Bot France 73: 270–288

Venkataramanan S V 1967 Spiral grain in chir (Pinus roxburghii Sargent) Proc IUFRO 14 Congr München IX Sect 41: 484–497

Vintoniv I S 1981 (Some physical and mechanical properties of wavy-grain wood of sycamore maple). Lesnoi Zh 6: 56–58

Vité J P 1958 Über die transpirationsphysiologische Bedeutung des Drehwuches bei Nadelhölzern. Forstwiss Cbl 77: 193–203

Vité J P 1959 Observations on the movement of injected dyes in Pinus ponderosa and Abies concolor. Contr Boyce Thompson Inst 20: 7–26

Vité J P 1967 Water conduction and spiral grain: causes and effects. Proc IUFRO 14 Congr München IX Sect 41: 338–351

Vité J P, Hendrickson W H 1958 Die Orientierung der Holzfasern im Douglasienstamm. Alg Forstz 14: 166–168

Vité J P, Rudinsky J A 1959 The water-conducting systems in conifers and their importance to the distribution of trunk injected chemicals. Contr Boyce Thompson Inst 20: 27–38

Walters C S 1951 Figured walnut propagated by grafting. J For 49: 917

Wareing P F, Hanney C E A, Digby J 1964 The role of endogenous hormones in cambial activity and xylem differentiation. In: Zimmermann M H (ed) The formation of wood in forest trees. Academic Press, New York, 323–346

Webb C D 1967 Interlocked grain in Liquidambar styraciflua. Proc IUFRO 14 Congr München IX Sect 41: 398–412

Webb C D 1969 Variation of interlocked grain in sweetgum. For Prod J 19(8): 45–48

Weddell E 1960 A study of the influence of "interlocked grain" on the bending strength of timber with particular reference to utile (Entandrophragma utile) and greenheart (Ocotea rodiaeri). J Timber Dev Assoc, London Res Report E/RR/14 Also: J Inst Wood Sci 7 (1961): 56–72

Wellner C A 1960 Devices to measure spiral grain and twisting in poles. J For 58: 117–118

Wellner C A, Lowery D P 1967 Spiral grain – a cause of pole twisting. US For Serv Res Pap INT-38, 17 pp

Wellwood R W, Smith J H G 1962 Variation in some important qualities of wood from young Douglas fir and hemlock trees. Fac For Univ of Brit Columbia, Vancouver, Res Pap 50, 15 pp

Wengert E M, Skaar C 1978 Additional consideration in measuring transverse moisture conductivity in wood. Wood Sci 11: 102–104

Wenham N W, Cusick F 1975 The growth of secondary wood fibres. New Phytol 74: 247–261

Wentworth C K 1931 Twist in the grain of coniferous trees. Science 73 (1885): 192

Whalley B E 1950 Increase in girth of the cambium in Thuja occidentalis L. Can J Res C 28: 331–340

Wheeler R T 1964 Equipment suitable for measuring angle of grain and for obtaining small timber samples from living trees. Comm For Rev 43: 314–318

Whyte A G D, Wiggins P C, Wong T W 1980 Spiral grain in Fijian grown P. caribaea var. hondurensis and its effects on sawn outturn. Proc IUFRO Div 5 Oxford, 18 pp

Wiggins P C 1979 Spiral grain in Pinus caribaea in Fiji. Thesis Univ Canterbury NZ, 167 pp

Wilcox M D 1978 Some problems in selecting to improve wood properties. In: Nikles D G, Burley J, Barnes R D (eds) Progress and problems of genetic improvement of tropical trees 1: 213–225
Wilson T R C 1921 The effect of spiral grain on the strength of wood. J For 19: 740–747
Włoch W 1976 Cell events in cambium connected with the formation and existence of a whirled cell arrangement. Acta Soc Bot Pol 45: 313–326
Włoch W 1981 Nonparallelism of cambium cells in neighbouring rows. Acta Soc Bot Pol 50: 625–636
Włoch W 1985 Time-variable frequency of events in domains of Tilia cambium. Acta Soc Bot Pol 54: 29–40
Włoch W, Bilczewska E 1987 Fibrillation of events in the cambial domains of Tilia cordata Mill. Acta Soc Bot Pol 56: 19–35
Włoch W, Zagórska-Marek B 1982 Reconstruction of storeyed cambium in the linden. Acta Soc Bot Pol 51: 215–228
Wodzicki T J 1980 Oscylacyjno-falowy mechanizm polarnego transportu auksyny a morfogeneza roslin. (Oscillatory-wavy mechanism of auxin polar transport and plant morphogenesis). Proc Second All-Polish Conf – Mechanisms of plant morphogenesis regulation, Rogow: 68–74
Wodzicki T J, Knegt E, Wodzicki A B, Bruinsma J 1984 Is indolyl-3-acetic acid involved in the wave-like pattern of auxin efflux from Pinus sylvestris stem segments? Physiol Plant 61: 209–213
Wodzicki T J, Wodzicki A B 1973 Auxin stimulation of cambial activity in Pinus silvestris II. Dependence on basipetal transport. Physiol Plant 29: 288–292
Wodzicki T J, Wodzicki A B 1981 Modulation of the oscillatory system involved in polar transport of auxin by other phytohormones. Physiol Plant 53: 176–180
Wodzicki T J, Wodzicki A B, Zajaczkowski S 1979 Hormonal modulation of the oscillatory system involved in polar transport of auxin. Physiol Plant 46: 97–100
Wodzicki T J, Abe H, Wodzicki A B, Pharis R P, Cohen J D 1987 Investigations on the nature of the auxin-wave in the cambial region of pine stems. Validation of IAA as the auxin component by the Avena coleoptile curvature assay and by gas chromatography-mass spectrometry selected ion monitoring. Plant Physiol 84: 135–143
Wong T W 1980 Twist in timber sawn from Pinus caribaea with a range of spiral grain angles in Fiji. Thesis Univ Canterbury NZ, 151 pp
Woodfin R O 1969 Spiral grain patterns in coast Douglas-fir. For Prod J 19(1): 53–60
Worrall J G 1980 The impact of environment on cambial growth. In: Control of shoot growth in forest trees. IUFRO Workshop, Fredricton, Canada, 127–142
Yamamoto H, Takahashi M, Kawaguchi N, Takizawa T 1978 (The mechanical restraint of distortion in structural lumber from plantation-grown Japanese larch 1. The twist restraint in 10.5 cm by 10.5 cm green square lumber with pith). J Hokkaido For Prod Res Inst 317: 7–11
Yeager W C 1931 Regarding twist in conifers. Science 73 (1893): 392–393
Ylinen A 1953 Über die mechanische Schaftformtheorie der Bäume. Holz Roh-Werkst 11: 209–210
Zagórska-Marek B 1975 Growth activity of fusiform initials in storeyed cambium. Acta Soc Bot Pol 44: 537–551
Zagórska-Marek B 1977 Zyawiska rozwoyowe w kambium pietrowym. (Developmental changes in storeyed cambium). Ph D Thesis, Wroclaw University, 43–60
Zagórska-Marek B 1980 Amplituda fali morfogenetycznej jako czynnik mogacy miec wplyw na charakter zmian falistej struktury kambium i typ Wloknistosci w drewnie. (Morphogenic wave amplitude and its effect on changes in cambial wavy structure and on the type of wavy grain in wood). Proc Second All-Polish Conf – Mechanism of plant morphogenesis regulation, Rogow, 147–151
Zagórska-Marek B 1984a Pseudotransverse divisions and intrusive elongation of fusiform initials in the storeyed cambium of Tilia. Can J Bot 62: 20–27
Zagórska-Marek B 1984b Microfibril orientation in figured wood. Presented to IAWA Tsukuba Japan, 3 pp
Zagórska-Marek B 1985 Domain dislocations in cambium of Fraxinus excelsior. Presented to IAWA Gainesville Florida, 1 p
Zagórska-Marek B, Hejnowicz Z 1980 Discontinuous lines on the radial face of wavy-grained xylem as a manifestation of morphogenetic waves in the cambium. Acta Soc Bot Pol 49: 49–62
Zagórska-Marek B, Little C H A 1986 Control of fusiform initial orientation in the vascular cambium of Abies balsamea stems by indol-3-ylacetic acid. Can J Bot 64: 1120–1128

Zajączkowski S 1980 Funkcjonowanie pola morfogenetycznego zwiazenego z polarnym transportem auksyny. (Functioning of the morphogentic field related to polar transport of auxin). Proc Second All-Polish Conf – Mechanisms of plant morphogenesis regulation, Warsaw Agr Univ, Rogow, 75–78

Zajączkowski S, Wodzicki T J 1978a On the question of stem polarity with respect to auxin transport. Physiol Plant 44: 122–126

Zajączkowski S, Wodzicki T J 1978b Auxin and plant morphogenesis – a model of regulation. Acta Soc Bot Pol 47: 233–243

Zajączkowski S, Wodzicki T J, Bruinsma 1983 A possible mechanism for whole-plant morphogenesis. Physiol Plant 57: 306–310

Zajączkowski S, Wodzicki T J, Romberger J A 1984 Auxin waves and plant morphogenesis. In: Scott T K (ed) Encyclopedia of plant physiology New Series Vol 10. Springer, Berlin, Heidelberg, New York, Tokyo, 244–262

Zakrewski J 1983 Hormonal control of cambial activity and vessel differentiation in Quercus robur. Physiol Plant 57: 537–542

Zalasky H 1972 Isolation and characteristics of sclerid-like cells in sapwood of Pinus and Populus. For Res Centre Can Infor Report NOR-X-48, 14 pp

Zobel B J 1961 Inheritance of wood properties in conifers. Silvae Genet 10: 65–70

Zobel B J 1964 Breeding for wood properties in forest trees. Unasylva 18: 89–103

Zobel B J 1965 Inheritance of spiral grain. Proc IUFRO Melbourne Sect 41, Vol 1, 9 pp

Zobel B J, Stonecypher R W, Browne C 1968 Inheritance of spiral grain in young loblolly pine. For Sci 14: 376–379

Subject Index

Abbreviations

ABA	Abscisic acid	IBA	Indolebutyric acid	SP	Pattern of spirality (within a stem)
AG	Angle of Grain	IG	Interlocked grain	TIBA	Triiodobenzoic acid
GA	Giberellic acid	LH	Left hand	WG	Wavy (curly) grain
IAA	Indoleacetic acid	RH	Right hand	ZEA	Zeatin
		SG	Spiral grain		

Springer Series in Wood Science

Editor: T. E. Timell

M. H. Zimmermann
Xylem Structure and the Ascent of Sap (1983)

J. F. Siau
Transport Processes in Wood (1984)

R. R. Archer
Growth Stresses and Strains in Trees (1986)

W. E. Hillis
Heartwood and Tree Exudates (1987)

S. Carlquist
Comparative Wood Anatomy (1988)

L. W. Roberts / P. B. Gahan / R. Aloni
Vascular Differentiation and Plant Growth Regulators (1988)

C. Skaar
Wood-Water Relations (1988)

J. M. Harris
Spiral Grain and Wave Phenomena in Wood Formation (1989)